T0137967

Lecture Notes in Statistics **146**

Springer
New York
Berlin
Heidelberg
Barcelona
Hong Kong
London
Milan
Paris
Singapore
Tokyo

Wim Schoutens

Stochastic Processes and Orthogonal Polynomials

Wim Schoutens
Departement Wiskunde
Faculteit Wetenschappen
Katholieke Universiteit Leuven
Celestijnenlaan 200 B
B-3001 Heverlee
Belgium

Library of Congress Cataloging-in-Publication Data
Schoutens, Wim.
Stochastic processes and orthogonal polynomials / Wim Schoutens.
p. cm.—(Lecture notes in statistics; 146)
Includes bibliographical references and index.
ISBN 0-387-95015-X (softcover: alk. paper)
1. Stochastic processes. 2. Orthogonal polynomials. I. Title.
II. Lecture notes in statistics (Springer-Verlag); v. 146.
QA274.S388 2000
519.2—dc21 00-022019

Printed on acid-free paper.

Camera-ready copy provided by the author.
Printed and bound by Sheridan Books, Ann Arbor, MI.
Printed in the United States of America.

9 8 7 6 5 4 3 2 1

ISBN 0-387-95015-X Springer-Verlag New York Berlin Heidelberg SPIN 10760181

To Ethel and Jente.

Preface

It has been known for a long time that there is a close connection between stochastic processes and orthogonal polynomials. For example, N. Wiener [112] and K. Ito [56] knew that Hermite polynomials play an important role in the integration theory with respect to Brownian motion. In the 1950s D.G. Kendall [66], W. Ledermann and G.E.H. Reuter [67] [74], and S. Karlin and J.L. McGregor [59] established another important connection. They expressed the transition probabilities of a birth and death process by means of a spectral representation, the so-called Karlin–McGregor representation, in terms of orthogonal polynomials. In the following years these relationships were developed further. Many birth and death models were related to specific orthogonal polynomials. H. Ogura [87], in 1972, and D.D. Engel [45], in 1982, found an integral relation between the Poisson process and the Charlier polynomials. Some people clearly felt the potential importance of orthogonal polynomials in probability theory. For example, P. Diaconis and S. Zabell [29] related Stein equations for some well-known distributions, including Pearson's class, with the corresponding orthogonal polynomials.

The most important orthogonal polynomials are brought together in the so-called Askey scheme of orthogonal polynomials. This scheme classifies the hypergeometric orthogonal polynomials that satisfy some type of differential or difference equation and stresses the limit relations between them.

In this work we

(1) introduce a link between the broad class of Sheffer polynomials and Lévy processes or sums of i.i.d. variables,

(2) show that the Krawtchouk polynomials play an important role, similar to the Hermite–Brownian and the Charlier–Poisson case, in stochastic integration (summation) theory with respect to the binomial process,

(3) bring the classical orthogonal polynomials into relationship with Stein's method for Pearson's and Ord's classes of distributions,

(4) give a chaotic representation for a general Lévy process, and finally

(5) we recall results concerning Karlin and McGregor's relation and extend these by considering doubly limiting conditional distributions. In this way we give a probabilistic interpretation of the major part of the Askey scheme.

In Chapter 1 we introduce the reader to the Askey scheme and orthogonal polynomials in general. Readers familiar with this subject can easily skip this chapter. We tried as much as possible to follow the accepted notation as presented, for example, in the Koekoek and Swarttouw reports [70] and [71].

In Chapter 2 all the main stochastic processes we need in this work are treated. All of them are Markov processes. For discrete time processes we look at Markov chains and random walks in particular. For continuous-time processes, we consider birth and death, Lévy, and diffusion processes.

The third chapter discusses the relationship between birth and death processes and orthogonal polynomials. The Karlin and McGregor representation for transition probabilities plays a central role. First we look at some birth and death processes, for example, the $M/M/\infty$ queue, the linear birth and death process, a quadratic model, and the Ehrenfest model. All of these processes are related to polynomials, called birth–death polynomials, of the Askey scheme. Next we extend the results of M. Kijima, M.G. Nair, P.K. Pollett, and E.A. van Doorn [68], who represented the conditional limiting distribution of a birth and death process in terms of the birth–death polynomials. We give a similar result for the doubly limiting conditional distribution. Furthermore similar results are also given for the discrete analogue of a birth and death process, the random walk.

An important chapter is the fourth one, where a new link is established between Sheffer polynomials and Lévy processes. The basic connection is a martingale relation. In a historic paper Meixner [81] determined all sets of orthogonal polynomials that are Sheffer polynomials. He proved that there were essentially five orthogonal Sheffer polynomial systems: the Hermite, Charlier, Meixner, Laguerre and the Meixner–Pollaczek polynomials. It turns out that for these polynomials, the measures (or distributions) of orthogonality are infinitely divisible and in this way we can associate some Lévy processes with them. The beautiful thing is that the monic polynomials, with suitable parameters, evaluated in its Lévy process are martingales. Besides the orthogonal Sheffer polynomials we also relate some other Sheffer polynomials to some Lévy processes. Sometimes the related distribution is not infinitely divisible; then we just consider the discrete counterparts: sums of i.i.d. random variables.

As mentioned above, the Hermite and Charlier polynomials play a major role in stochastic integration theory with respect to Brownian motion and the Poisson process, respectively. In fact they are the stochastic counterparts of the monomials of classical deterministic integration theory. Integrating the polynomial of degree n evaluated in the process with respect to the compensated process, gives the polynomial of degree $n+1$. In Chapter 5, we give a similar result for the binomial process and the Krawtchouk polynomials. Note that the involved martingale relations of Chapter 4 are an implicit result of these integrals. Furthermore by taking generating functions we obtain the stochastic counterparts of the exponential function. In the Brownian motion case, this stochastic exponential, often called geometric Brownian motion, plays an important role in the celebrated Black–Scholes option pricing model [21]. We give the stochastic exponential for the Poisson and binomial processes. Furthermore, an important part of Chapter 6 discusses the chaotic representation property and the weaker predictable representation property. For a general Lévy process (satisfying some moment conditions), we introduce power jump processes and the related Teugels martingales. In addition we orthogonalize the Teugels martingales and show how their orthogonalization is closely related to classical orthogonal polynomials. We give a chaotic representation for every square integral random variable in terms of these orthogonalized Teugels martingales. The predictable representation with respect to the same set of orthogonalized martingales of square integrable random variables and of square integrable martingales is an easy consequence of chaotic representation.

In the last chapter, Chapter 6, we bring the classical orthogonal polynomials into relationship with Stein's method for Pearson's and Ord's class of distributions. P. Diaconis and S. Zabell already mentioned this connection. We give a systematic treatment, including the Stein equation, its solution, and smoothness conditions. A key tool in Stein's theory is the generator method developed by Barbour, who suggested employing the generator of a Markov process for the operator of the Stein equation. It turns out that the transition probabilities of these Markov processes, in our case a birth and death process for discrete distributions and a diffusion process for continuous distributions, can be given by means of a spectral representation in terms of the related orthogonal polynomials.

Some of the important ingredients of distributions, orthogonal polynomials and Sheffer systems, together with some duality relations are summarized in the appendices.

I owe special thanks to my scientific mentor J.L. Teugels for his continuous interest and encouragement. I am also extremely grateful to the other members of my doctoral jury Walter Van Assche, Arno Kuijlaars, Nick Bingham, Erik van Doorn, and Adhemar Bultheel. It is a pleasure to thank David Nualart for his hospitality and support. I also want to thank the Fund for Scientific Research — Flanders (Belgium), the K. U. Leuven, and the EURANDOM Institute for their support. And, last and most, I want to thank my wife, Ethel, and son, Jente, for their love.

W. Schoutens
Leuven

Contents

1
The Askey Scheme of Orthogonal Polynomials

The main focus of this book is the relationship between orthogonal polynomials and stochastic processes. In this chapter we review the relevant background of orthogonal polynomials. We start with some preliminaries and introduce the concept of an orthogonal polynomial. After classifying the so-called classical orthogonal polynomials, we describe the Askey scheme. This scheme contains the most important orthogonal polynomials and stresses the limit relations between them.

1.1 Preliminaries

Pochhammer's Symbol

Throughout this work we find it convenient to employ the *Pochhammer symbol* $(a)_n$ defined by

$$(a)_n = \begin{cases} 1, & \text{if } n = 0, \\ a(a+1)\dots(a+n-1), & \text{if } n = 1, 2, 3, \dots. \end{cases}$$

In terms of Gamma functions, we have

$$(a)_n = \frac{\Gamma(a+n)}{\Gamma(a)}, \quad n > 0.$$

The Generalized Hypergeometric Series

The *generalized hypergeometric series* ${}_pF_q$ is defined by

$$ {}_pF_q(a_1,\ldots,a_p;b_1,\ldots,b_q;z) = \sum_{j=0}^{\infty} \frac{(a_1)_j\cdots(a_p)_j}{(b_1)_j\cdots(b_q)_j}\frac{z^j}{j!}, $$

where $b_i \neq 0,-1,-2,\ldots$, $i=1,\ldots,q$. There are p numerator parameters and q denominator parameters. Clearly, the orderings of the numerator parameters and of the denominator parameters are immaterial.

The cases ${}_0F_0$ and ${}_1F_0$ are elementary: exponential, respectively, binomial series. The case ${}_2F_1$ is the familiar Gauss hypergeometric series.

If one of the numerator parameters a_i, $i=1,\ldots,p$ is a negative integer, $a_1=-n$, say, the series terminates and

$$ {}_pF_q(-n,\ldots,a_p;b_1,\ldots,b_q;z) = \sum_{j=0}^{n} \frac{(-n)_j\cdots(a_p)_j}{(b_1)_j\cdots(b_q)_j}\frac{z^j}{j!}. $$

When the series is infinite, it converges for $|z|<\infty$ if $p\leq q$, it converges for $|z|<1$ if $p=q+1$, and it diverges for all $z\neq 0$ if $p>q+1$.

Finally, we introduce a notation for the Nth partial sum of a generalized hypergeometric series, which is particularly useful if one of the b_j equals $-N$. We use this notation in the definition of the discrete orthogonal polynomials. We define

$$ {}_p\tilde{F}_q(a_1,\ldots,a_p;b_1,\ldots,b_q;z) = \sum_{j=0}^{N} \frac{(a_1)_j\cdots(a_p)_j}{(b_1)_j\cdots(b_q)_j}\frac{z^j}{j!}, $$

where N denotes the nonnegative integer that appears in some definitions of a family of discrete orthogonal polynomials.

Monic Version of a Polynomial

For a general polynomial of degree n,

$$ Q_n(x) = a_nx^n + a_{n-1}x^{n-1} + \ldots + a_0, \quad a_n \neq 0 $$

we call a_n the *leading coefficient* of the polynomial and we denote by

$$ \tilde{Q}_n(x) = \frac{Q_n(x)}{a_n} = x^n + \frac{a_{n-1}}{a_n}x^{n-1} + \ldots + \frac{a_0}{a_n} $$

the *monic version* of this polynomial, i.e., with the leading coefficient equal to one.

1.2 Orthogonal Polynomials

1.2.1 Orthogonality Relations

A system of polynomials $\{Q_n(x), n \in \mathcal{N}\}$ where $Q_n(x)$ is a polynomial of exact degree n and $\mathcal{N} = \mathbb{N} = \{0, 1, 2, \ldots\}$ or $\mathcal{N} = \{0, 1, \ldots, N\}$ for a finite nonnegative integer N, is *an orthogonal system of polynomials with respect to some real positive measure* ϕ, if we have the following orthogonality relations

$$\int_S Q_n(x)Q_m(x)d\phi(x) = d_n^2\delta_{nm}, \quad n, m \in \mathcal{N}, \tag{1.1}$$

where S is the support of the measure ϕ and the d_n are nonzero constants. If these constants $d_n = 1$, we say the system is *orthonormal* .

The measure ϕ usually has a density $\rho(x)$ or is a discrete measure with weights $\rho(i)$ at the points x_i. The relations (1.1) then become

$$\int_S Q_n(x)Q_m(x)\rho(x)dx = d_n^2\delta_{nm}, \quad n, m \in \mathcal{N}, \tag{1.2}$$

in the former case and

$$\sum_{i=0}^{M} Q_n(x_i)Q_m(x_i)\rho_i = d_n^2\delta_{nm}, \quad n, m \in \mathcal{N}, \tag{1.3}$$

in the latter case where it is possible that $M = \infty$.

1.2.2 Three-Term Recurrence Relation

It is well known that all orthogonal polynomials $\{Q_n(x)\}$ on the real line satisfy a three-term recurrence relation

$$-xQ_n(x) = b_nQ_{n+1}(x) + \gamma_nQ_n(x) + c_nQ_{n-1}(x), \quad n \geq 1, \tag{1.4}$$

where $b_n, c_n \neq 0$ and $c_n/b_{n-1} > 0$. Note that if for all $n \in \{0, 1, \ldots\}$,

$$Q_n(0) = 1, \tag{1.5}$$

we have $\gamma_n = -(b_n + c_n)$ and the polynomials $Q_n(x)$ can be defined by the recurrence relation

$$-xQ_n(x) = b_nQ_{n+1}(x) - (b_n + c_n)Q_n(x) + c_nQ_{n-1}(x).$$

together with $Q_{-1}(x) = 0$ and $Q_0(x) = 1$. Favard proved a converse result [25].

Theorem 1 (Favard's Theorem) *Let $A_n, B_n,$ and C_n be arbitrary sequences of real numbers and let $\{Q_n(x)\}$ be defined by the recurrence relation*

$$Q_{n+1}(x) = (A_nx + B_n)Q_n(x) - C_nQ_{n-1}(x), \quad n \geq 0,$$

together with $Q_0(x) = 1$ *and* $Q_{-1}(x) = 0$. *Then the* $\{Q_n(x)\}$ *are a system of orthogonal polynomials if and only if* $A_n \neq 0$, $C_n \neq 0$, *and* $C_n A_n A_{n-1} > 0$ *for all* n.

If all $A_n = 1$ we have a system of monic orthogonal polynomials.

1.3 Classical Orthogonal Polynomials

1.3.1 Hypergeometric Type Equations

Hypergeometric Type Differential Equations

Many problems of applied mathematics, and theoretical and mathematical physics lead to equations of the form

$$s(x)y'' + \tau(x)y' + \lambda y = 0, \tag{1.6}$$

where $s(x)$ and $\tau(x)$ are polynomials of at most second and first degree, respectively, and λ is a constant. We refer to (1.6) as *a differential equation of hypergeometric type*, and its solutions as *functions of hypergeometric type*.

If furthermore

$$\lambda = \lambda_n = -n\tau' - \frac{1}{2}n(n-1)s'',$$

the equation of hypergeometric type has a particular solution of the form $y(x) = y_n(x)$ which is a polynomial of degree n. We call such solutions *polynomials of hypergeometric type*. The polynomials $y_n(x)$ are the simplest solutions of (1.6).

In Section 1.3.2 we classify all possible solutions, up to affine transformations. It can be shown [84] that these polynomial solutions of (1.6) have the orthogonality property

$$\int_a^b y_m(x) y_n(x) \rho(x) dx = d_n^2 \delta_{nm},$$

for some constants a, b possibly infinite, $d_n \neq 0$, and where the weight function of orthogonality $\rho(x)$ satisfies the differential equation

$$(s(x)\rho(x))' = \tau(x)\rho(x). \tag{1.7}$$

These polynomials of hypergeometric type $y_n(x)$ are known as the *(very) classical orthogonal polynomials of a continuous variable* .

In Chapter 7, Equation (1.7) is the starting point for the discussion of Stein's method for Pearson's class of distributions. This class covers probability distributions satisfying this type of differential equation.

We note in conclusion that the system of classical orthogonal polynomials $\{y_n(x)\}$ is *complete* on the interval $S = (a, b)$ for functions in $L^2(S, \rho)$, i.e., for functions $f(x)$ that satisfy the condition of square integrability

$$\int_a^b f^2(x)\rho(x)dx < \infty;$$

i.e., each $f \in L^2(S, \rho)$ has a Fourier expansion of the form

$$f(x) = \sum_{n=0}^{\infty} a_n y_n(x), \quad x \in S,$$

where

$$a_n = \frac{\int_a^b f(x)y_n(x)\rho(x)dx}{d_n^2}, \quad n = 0, 1, \ldots,$$

and convergence is in the L^2-sense.

Hypergeometric Type Difference Equation

First we introduce some notation. We set

$$\Delta f(x) = f(x+1) - f(x) \quad \text{and} \quad \nabla f(x) = f(x) - f(x-1). \tag{1.8}$$

A difference equation of hypergeometric type is one of the form

$$s(x)\Delta\nabla y(x) + \tau(x)\Delta y(x) + \lambda y(x) = 0, \tag{1.9}$$

where $s(x)$ and $\tau(x)$ are polynomials of at most second and first degree, respectively, and λ is a constant. Using (1.8) we can rewrite (1.9) as

$$(s(x)+\tau(x))y(x+1) - (2s(x)+\tau(x))y(x) + s(x)y(x-1) = -\lambda y(x). \tag{1.10}$$

If again

$$\lambda = \lambda_n = -n\tau' - \frac{1}{2}n(n-1)s'',$$

the difference equation of hypergeometric type has a particular solution of the form $y(x) = y_n(x)$ which is a polynomial of degree n, provided

$$\mu_m = \lambda + m\tau' + \frac{1}{2}m(m-1)s'' \neq 0$$

for $m = 0, 1, \ldots, n-1$.

In Section 1.3.2 we classify all possible solutions, up to affine transformations. It can be shown [84] that the polynomial solutions of (1.9) have the orthogonality property:

$$\sum_{x=a}^{b} y_m(x)y_n(x)\rho(x) = d_n^2\delta_{nm},$$

for some constants a, b possibly infinite, $d_n \neq 0$, and where the discrete orthogonality measure $\rho(x)$ satisfies the difference equation

$$\Delta(s(x)\rho(x)) = \tau(x)\rho(x). \tag{1.11}$$

These polynomials of hypergeometric type $y_n(x)$ are known as the *classical orthogonal polynomials of a discrete variable.* In Chapter 7, Equation (1.11) is the starting point for the discussion of Stein's method for Ord's class of distributions. This class covers all probability distributions satisfying the difference equation (1.11).

1.3.2 Classification of Classical Orthogonal Polynomials

For exact definitions and a summary of the key ingredients of classical orthogonal polynomials we refer the reader to Appendix A.

Classical Orthogonal Polynomials of a Continuous Variable

There are in essence five basic solutions of (1.7), depending on whether the polynomial $s(x)$ is constant, linear, or quadratic and, in the last case, on whether the discriminant $D = b^2 - 4ac$ of $s(x) = ax^2 + bx + c$ is positive, negative, or zero.

1. *Jacobi,* $\deg s(x) = 2, D > 0$. If we take $s(x) = 1 - x^2$ and $\tau(x) = -(\alpha+\beta+2)x+\beta-\alpha$, then $\rho(x) = (1-x)^\alpha(1+x)^\beta$, the *Beta kernel.* Furthermore, $\lambda_n = n(n+\alpha+\beta+1)$ and the corresponding polynomials are called the *Jacobi polynomials* and are denoted by $P_n^{(\alpha,\beta)}(x)$. For the Jacobi polynomials the orthogonality relation will be satisfied if $a = -1$, $b = 1$, $\alpha, \beta > -1$.

2. *Bessel,* $\deg s(x) = 2, D = 0$. After an affine change of variable, $\rho(x)$ can be brought into the form $\rho(x) = Cx^{-\alpha}\exp(-\beta/x)$, where C is the appropriate normalizing constant. If it is assumed that $\rho(x)$ is defined on $(0, +\infty)$, $\alpha > 1$ and $\beta \geq 0$ ensure that $\rho(x)$ is integrable. In this case $s(x) = x^2$ and $\tau(x) = (2-\alpha)x + \beta$. Because for this distribution only the moments of order strictly less than $\alpha - 1$ exist, it is impossible to have an infinite system of orthogonal polynomials.

3. *Romanovski,* $\deg s(x) = 2, D < 0$. After an affine change of variable, $\rho(x)$ can be brought into the form

   $$\rho(x) = C(1+x^2)^{-\alpha}\exp(\beta\arctan(x)),$$

   where C is the appropriate normalizing constant. If it is assumed that $\rho(x)$ is defined on $(-\infty, +\infty)$, then $\alpha > 1/2$ and $\beta \in \mathbb{R}$. In particular for $\alpha = (n+1)/2$ and $\beta = 0$, $s(x) = 1 + x^2/n$ and $\tau(x) = -(n-1)x/n$, $n \in \{1, 2, \ldots\}$; then $\rho(x) = C(1 + x^2/n)^{-(n+1)/2}$, with $C =$

$\Gamma((n+1)/2)/(\sqrt{n\pi}\Gamma(n/2))$. This gives the Student's t-distribution t_n. Because $\lambda_n = 0$ we do not have an infinite system of orthogonal polynomials.

4. *Laguerre*, $\deg s(x) = 1$. Let $s(x) = x$ and $\tau(x) = -x + \alpha + 1$; then $\rho(x) = x^{\alpha}e^{-x}/\Gamma(\alpha+1)$, the density of the *Gamma distribution* $\mathrm{G}(\alpha, 1)$. Furthermore $\lambda_n = n$ and the polynomials $y_n(x)$ are called the *Laguerre polynomials* and are denoted by $L_n^{(\alpha)}(x)$. The Laguerre polynomials will satisfy the orthogonality relation when $a = 0$, $b = \infty$, and $\alpha > -1$.
5. *Hermite*, $\deg s(x) = 0$. Let $s(x) = 1$ and $\rho(x) = \exp(-x^2)/\sqrt{\pi}$, a *normal distribution* $\mathrm{N}(0, 1/2)$. Then $\tau(x) = -2x$ and $\lambda_n = 2n$. The polynomials $y_n(x)$ are called the *Hermite polynomials* and are denoted by $H_n(x)$. Hermite polynomials are orthogonal on the interval $(-\infty, \infty)$. We work here mostly with the rescaled Hermite polynomials $H_n(x/\sqrt{2})$ which are orthogonal with respect to the *standard normal distribution* $\mathrm{N}(0,1)$ with density function given by $\rho(x) = \exp(-x^2/2)/\sqrt{2\pi}$.

Although there are some arguments for including the Bessel and Romanovski polynomials [75] into the Askey scheme, they are not a part of it.

Orthogonal Polynomials of a Discrete Variable

In order to find explicit expressions for $\rho(x)$ we rewrite the difference equation (1.11) in the form

$$\frac{\rho(x+1)}{\rho(x)} = \frac{s(x)+\tau(x)}{s(x+1)}. \tag{1.12}$$

Let $s(x)$ be a polynomial of second degree. After some calculations [84] one finds that all the solutions $y_n(x)$ can be transformed into the *Hahn polynomials* $Q_n(x; \alpha, \beta, N)$. They are defined by

$$Q_n(x; \alpha, \beta, N) = {}_3\tilde{F}_2(-n, n+\alpha+\beta+1, -x; \alpha+1, -N; 1),$$

and satisfy the three-term recurrence relation (1.4) with

$$\begin{aligned} b_n &= \frac{(n+\alpha+\beta+1)(n+\alpha+1)(N-n)}{(2n+\alpha+\beta+1)(2n+\alpha+\beta+2)}, \\ c_n &= \frac{n(n+\beta)(n+\alpha+\beta+N+1)}{(2n+\alpha+\beta)(2n+\alpha+\beta+1)}, \\ \gamma_n &= -(b_n + c_n). \end{aligned} \tag{1.13}$$

Hahn polynomials satisfy orthogonality relations

$$\sum_{x=0}^{N} Q_n(x)Q_m(x)\rho(x) = d_n^2\delta_{nm}, \quad n, m = 0, 1, \ldots, N,$$

where

$$\rho(x) = \binom{N}{x} \frac{(\alpha+1)_x(\beta+1)_{N-x}}{(\alpha+\beta+2)_N}, \quad x = 0, 1, \ldots, N, \tag{1.14}$$

is the *hypergeometric distribution* $\mathrm{HypI}(\alpha, \beta, N)$, and

$$\pi_n = 1/d_n^2 = \binom{N}{n} \frac{2n+\alpha+\beta+1}{\alpha+\beta+1} \frac{(\alpha+1)_n(\alpha+\beta+1)_n}{(\beta+1)_n(N+\alpha+\beta+2)_n}. \tag{1.15}$$

For these Hahn polynomials we have $s(x) = x(N - x + \beta + 1)$, $\tau(x) = N(\alpha+1) - x(\alpha+\beta+2)$, and $\lambda_n = n(n+\alpha+\beta+1)$.

If we set $\alpha = -\tilde{\alpha} - 1$ and $\beta = -\tilde{\beta} - 1$, then we obtain

$$\tilde{\rho}(x) = \frac{\binom{\tilde{\alpha}}{x}\binom{\tilde{\beta}}{N-x}}{\binom{\tilde{\alpha}+\tilde{\beta}}{N}}, \quad x = 0, 1, \ldots, N,$$

which is the *hypergeometric distribution* $\mathrm{HypII}(\tilde{\alpha}, \tilde{\beta}, N)$.

Let $s(x)$ be a polynomial of the first degree. We consider three cases:

$$s(x) + \tau(x) = \begin{cases} m(\gamma + x) \\ m(\gamma - x) \\ m \end{cases}.$$

Here m and γ are positive constants. Then (1.12) has the following solutions

$$\rho(x) = \begin{cases} C(m/(m+1))^x(\gamma)_x/x! \\ Cm^x/(x!\Gamma(\gamma+1-x)) \\ Cm^x/x! \end{cases}.$$

1. In the first case we take

   $$\mu = m/(m+1), \quad a = 0, \quad b = \infty, \quad 0 < \mu < 1, \gamma > 0,$$

   and $C = (1-\mu)^\gamma$. We then obtain *the negative binomial distribution* or *Pascal distribution* $\mathrm{Pa}(\gamma, \mu)$,

   $$\rho(x) = (1-\mu)^\gamma \mu^x (\gamma)_x/x!.$$

   The corresponding polynomials are called the *Meixner polynomials* $M_n(x; \gamma, \mu)$.

2. In the second case, we take

   $$a = 0, b = N, \gamma = N, m = p/(1-p), 0 < p < 1, C = (1-p)^N N!.$$

   The numbers $\rho(x)$ become the familiar *binomial distribution* denoted by $\mathrm{Bin}(N, p)$,

   $$\rho(x) = \binom{N}{x} p^x q^{N-x},$$

   where $q = 1 - p$. The corresponding polynomials are called the *Krawtchouk polynomials* and are denoted by $K_n(x; N, p)$.

3. In the third case, with

$$a = 0, \quad b = +\infty, \quad m = \mu, \quad C = e^{-\mu},$$

we have the *Poisson distribution* $\mathrm{P}(\mu)$,

$$\rho(x) = e^{-\mu}\mu^x/x!.$$

The corresponding orthogonal polynomials are the *Charlier polynomials* $C_n(x;\mu)$.

Note that the Charlier polynomials are limit cases of the Krawtchouk and Meixner polynomials, which are themselves limit cases of the Hahn polynomials.

The case $s(x) = 1$ is not of interest, since it does not lead to any new polynomials.

Duality Relations

The Krawtchouk, Meixner, and Charlier polynomials are *self-dual*; i.e., they satisfy

$$p_n(x) = p_x(n), \tag{1.16}$$

for relevant x and n.

In this way the second-order difference equation (1.10)

$$(s(x)+\tau(x))y_n(x+1) - (2s(x)+\tau(x))y_n(x) + s(x)y_n(x-1) = -\lambda_n y_n(x)$$

becomes the three-term recurrence relation

$$(s(n)+\tau(n))y_{n+1}(x) - (2s(n)+\tau(n))y_n(x) + s(n)y_{n-1}(x) = -\lambda_x y_n(x)$$

after interchanging x and n.

Dual Hahn Polynomials

For the Hahn polynomials there is no self-duality. Define

$$R_n(x(x+\alpha+\beta+1);\alpha,\beta,N) = Q_x(n;\alpha,\beta,N), \quad 0 \le n, x \le N. \tag{1.17}$$

Then

$$R_n(x(x+\alpha+\beta+1);\alpha,\beta,N) = {}_3\tilde{F}_2(-n,-x,x+\alpha+\beta+1;\alpha+1,-N;1).$$

It can be shown that R_n is a polynomial of degree n in $x(x+\alpha+\beta+1)$. The R_n are called the *dual Hahn polynomials.* We have the following orthogonality relations.

$$\sum_{x=0}^{N} R_m(x(x+\alpha+\beta+1))R_n(x(x+\alpha+\beta+1))\pi_x = \delta_{mn}/\rho(n).$$

Here π_x and $\rho(n)$ are as in (1.15) and (1.14).

The three-term recurrence relation for Hahn polynomials translates into a second-order difference equation which is a slight generalization of (1.10). It has the form

$$a(x)p_n(\lambda(x+1))+b(x)p_n(\lambda(x))+c(x)p_n(\lambda(x-1)) = -\lambda_n p_n(\lambda(x)), \quad (1.18)$$

where in this case $\lambda(x) = x(x+\alpha+\beta+1)$ is a quadratic function of x, $\lambda_n = n$, $a(x) = b_x$, $b(x) = -(a(x)+c(x))$, and $c(x) = c_x$, where the b_x and c_x are the coefficients in the three-term recurrence relation of the Hahn polynomials given in (1.13).

Note that the Krawtchouk and Meixner polynomials are also limit cases of the dual Hahn polynomials.

1.4 The Askey Scheme

Racah Polynomials

It is a natural question to ask for other orthogonal polynomials to be eigenfunctions of a second-order difference equation of the form (1.18). A four-parameter family with this property is the *Racah polynomials*, essentially known for a long time to physicists as Racah coefficients. However, before the late 1970s [114] it was not recognized that orthogonal polynomials are hidden in the Racah coefficients. Racah polynomials are defined by

$$R_n(x(x+\gamma+\delta+1);\alpha,\beta,\gamma,\delta) = \quad (1.19)$$
$$_4\tilde{F}_3(-n, n+\alpha+\beta+1, -x, x+\gamma+\delta+1; \alpha+1, \beta+\delta+1, \gamma+1; 1),$$

where $\alpha+1$ or $\beta+\delta+1$ or $\gamma+1 = -N$ for some $N \in \{0,1,2,\ldots\}$ and where $n = 0,1,\ldots,N$.

It can be shown that the Racah polynomial R_n is indeed a polynomial of degree n in $\lambda(x) = x(x+\gamma+\delta+1)$. The Racah polynomials are orthogonal with respect to weights $\rho(x)$ on the points $\lambda(x)$, $x = 0,1,\ldots,N$, given by

$$\rho(x) = \frac{(\gamma+\delta+1)_x((\gamma+\delta+3)/2)_x(\alpha+1)_x(\beta+\delta+1)_x(\gamma+1)_x}{x!((\gamma+\delta+1)/2)_x(\gamma+\delta-\alpha+1)_x(\gamma-\beta+1)_x(\delta+1)_x}.$$

It is evident from (1.19) that dual Racah polynomials are again Racah polynomials with α, β interchanged with γ, δ. Hahn and dual Hahn polynomials can be obtained as limit cases of Racah polynomials.

Wilson Polynomials

Another system, closely related to the Racah polynomials, is the *Wilson polynomials* W_n [114] defined by

$$\frac{W_n(x^2;a,b,c,d)}{(a+b)_n(a+c)_n(a+d)_n} =$$

$$_4F_3(-n, n+a+b+c+d-1, a+ix, a-ix; a+b, a+c, a+d; 1).$$

Apparently, the right-hand side defines a polynomial of degree n in x^2. If $Re(a,b,c,d) > 0$ and nonreal parameters occur in conjugate pairs then the functions $W_n(x^2)$ are orthogonal with respect to $\rho(x)$ on $[0, +\infty)$, where

$$\rho(x) = \left| \frac{\Gamma(a+ix)\Gamma(b+ix)\Gamma(c+ix)\Gamma(d+ix)}{\Gamma(2ix)} \right|^2 .$$

Continuous Hahn and Continuous Dual Hahn Polynomials

Now, we can descend from the Wilson polynomials by limit transitions, just as from the Racah polynomials. On the $_3F_2$ level we thus get continuous analogues of the Hahn and the dual Hahn polynomials as follows.

The *continuous dual Hahn polynomials* S_n are given by

$$S_n(x^2; a, b, c) = (a+b)_n(a+c)_n \; {}_3F_2(-n, a+ix, a-ix; a+b, a+c; 1),$$

where a, b, c have positive real parts; if one of these parameters is not real then one of the other parameters is its complex conjugate. The functions $S_n(x^2)$ are orthogonal with respect to $\rho(x)$ on $[0, +\infty)$, where

$$\rho(x) = \left| \frac{\Gamma(a+ix)\Gamma(b+ix)\Gamma(c+ix)}{\Gamma(2ix)} \right|^2 .$$

The *continuous Hahn polynomials* p_n are given by

$$\frac{n! p_n(x; a, b, \bar{a}, \bar{b})}{(a+\bar{a})_n(a+\bar{b})_n i^n} = {}_3F_2(-n, n+a+\bar{a}+b+\bar{b}-1, a+ix; a+\bar{a}, a+\bar{b}; 1),$$

where a, b have a positive real part. The polynomials P_n are orthogonal on $\mathbb{R}$ with respect to the $\rho(x) = |\Gamma(a+ix)\Gamma(b+ix)|^2$.

Jacobi polynomials are limit cases of continuous Hahn polynomials and also directly of Wilson polynomials (with one pair of complex conjugate parameters).

Meixner–Pollaczek Polynomials

There is one further class of orthogonal polynomials on the $_2F_1$ level occurring as limit cases of orthogonal polynomials on the $_3F_2$ level: the *Meixner–Pollaczek polynomials*, defined by

$$P_n^{(a)}(x; \phi) = \frac{(2a)_n \exp(in\phi)}{n!} \; {}_2F_1(-n, a+ix; 2a; 1-e^{-2i\phi}),$$

where $a > 0$ and $0 < \phi < \pi$. They are orthogonal on $\mathbb{R}$ with respect to $\rho(x) = \exp(2\phi - \pi)|\Gamma(a+ix)|^2$ and are limits of both continuous Hahn and continuous dual Hahn polynomials. Laguerre polynomials are limit cases of Meixner–Pollaczek polynomials.

Askey Scheme

All families of orthogonal polynomials previously discussed, together with the limit transitions between them, form the *Askey scheme of hypergeometric orthogonal polynomials.* The exact limit relations between all these polynomials can be found in Appendix E. We summarize all limit relations in Figure 1.1.

Notes

For the history of the Askey scheme we refer to [3]. The scheme first appeared in 1985 in the appendix of [5], but on the third level it states the continuous symmetric Hahn polynomials instead of the more general continuous Hahn polynomials on the second level. Askey and Wilson missed this more general case and therefore included a level that was not necessary. The more general case (continuous Hahn polynomials) was found later by Atakishiev and Suslov. Essentially the same table, with the same "mistake," was published in [73]. There exists a generalization by Koekoek and Swartouw [70] [71] of the table including q-polynomials. Still, not all q-generalizations are known, especially the cases with $q > 1$, which are the subject of further investigations.

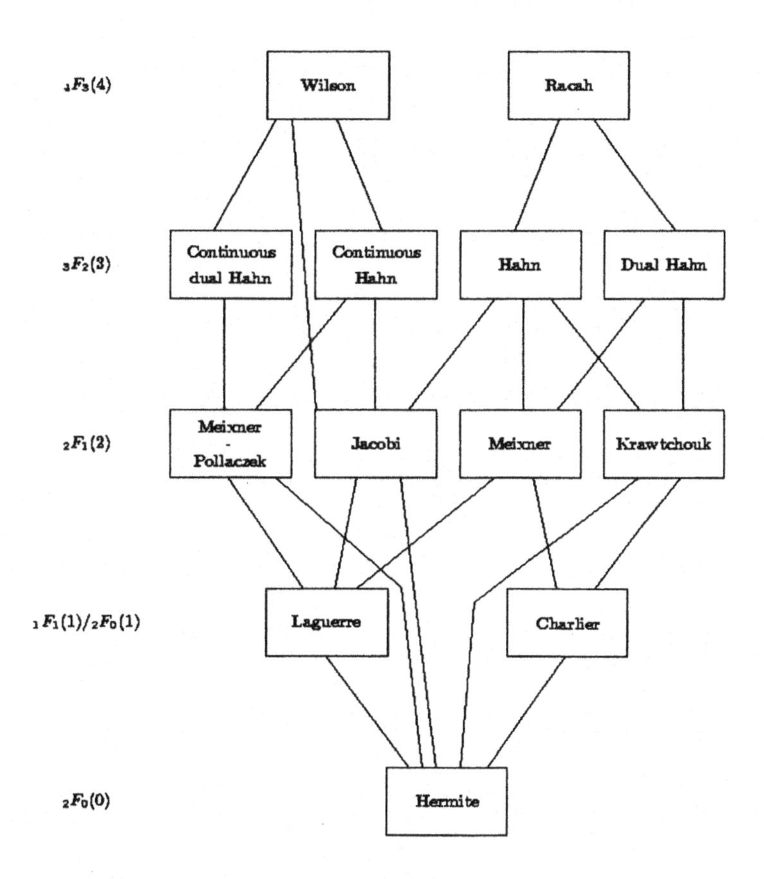

FIGURE 1.1. The Askey Scheme

2

Stochastic Processes

2.1 Markov Processes

Markov Property

One important type of stochastic process is a *Markov process*, a stochastic process that has a limited form of "historical" dependency. To precisely define this dependency, let $\{X_t, t \in \mathcal{T}\}$ be a stochastic process defined on the parameter set $\mathcal{T}$. We think of $\mathcal{T} \subset [0, \infty)$ in terms of time, and the values that X_t can assume are called the *states* which are elements of a *state space* $S \subset \mathbb{R}$. A stochastic process is called a Markov process if it satisfies

$$\begin{aligned} &\Pr(X_{t_0+t_1} \leq x | X_{t_0} = x_0, X_\tau, 0 \leq \tau < t_0) \\ &= \Pr(X_{t_0+t_1} \leq x | X_{t_0} = x_0), \end{aligned} \tag{2.1}$$

for any value of $t_0, t_1 > 0$. To interpret (2.1), we think of t_0 as being the present time. Equation (2.1) states that the evolution of a Markov process at a future time, conditioned on its present and past values, depends only on its present value. Expressed differently, the present value of X_{t_0} contains all the information about the past evolution of the process that is needed to determine the future distribution of the process. The condition (2.1) that defines a Markov process is sometimes termed the *Markov property*.

Markov processes are classified by whether the sets $\mathcal{T}$ and S are *discrete* (containing a countable number of elements) or *continuous* (containing an uncountable number of elements). Throughout this text a *Markov chain* has a discrete parameter set $\mathcal{T}$ and a discrete state space S.

Infinitesimal Generator

Suppose X_t is a continuous time Markov process on the state space S, having right-continuous sample paths and a transition density

$$p(t; x, y) \equiv \frac{\partial}{\partial y} \Pr(X_t \leq y | X_0 = x),$$

if S is continuous or transition probabilities

$$p(t; x, y) \equiv \Pr(X_t = y | X_0 = x),$$

if S is discrete.

On the set $\mathcal{B}(S)$ of all real-valued, bounded, Borel measurable functions f on S define the transition operator

$$(T_t f)(x) \equiv E[f(X_t)|X_0 = x], \quad t \geq 0.$$

Depending on whether S is continuous or discrete, we can rewrite this as

$$(T_t f)(x) = E[f(X_t)|X_0 = x] = \int_S f(y) p(t; x, y) dy, \quad t \geq 0$$

in the former case, and

$$(T_t f)(x) = E[f(X_t)|X_0 = x] = \sum_{y \in S} f(y) p(t; x, y), \quad t \geq 0$$

in the latter case. Then T_t is a bounded linear operator on $\mathcal{B}(S)$ when the latter is given the *sup norm* defined by

$$||f|| \equiv \sup\{|f(y)| : y \in S\}.$$

For all $f \in \mathcal{B}(S)$, T_t is a contraction; i.e., $||T_t f|| \leq ||f||$. Furthermore, the family of transition operators $\{T_t, t > 0\}$ has the *semigroup property*

$$T_{s+t} = T_s T_t,$$

where the right side denotes the composition of two maps. This relation also implies that the transition operators *commute*,

$$T_s T_t = T_t T_s.$$

In [17] it is shown that the derivative (operator) of the function $t \to T_t$ at $t = 0$ determines $\{T_t, t > 0\}$.

Definition 1 *The infinitesimal generator $\mathcal{A}$ of the family of operators $\{T_t, t > 0\}$, or of the Markov process X_t, is the linear operator $\mathcal{A}$ defined by*

$$(\mathcal{A}f)(x) = \lim_{s \downarrow 0} \frac{(T_s f)(x) - f(x)}{s},$$

for all $f \in \mathcal{B}(S)$ such that the right side converges to some function uniformly in x. The class of all such f is the domain $\mathcal{D}_{\mathcal{A}}$ of $\mathcal{A}$.

In the next sections we look more closely at some special types of Markov processes.

2.2 Markov Chains

A discrete time Markov chain $\{X_n\}$ is a Markov process whose state space is a countable or finite set and for which $\mathcal{T} = \{0, 1, \ldots\}$. It is frequently convenient to label the state space of the process by the nonnegative integers $S = \{0, 1, \ldots\}$, but sometimes an additional state -1 is included.

The probability of X_{n+1} being in state j, given that X_n is in state i, called a one-step transition probability, is denoted by $P_{ij}^{n,n+1}$; i.e.,

$$P_{ij}^{n,n+1} = \Pr(X_{n+1} = j | X_n = i), \quad i, j \in S.$$

The notation emphasizes that in general the transition probabilities are functions not only of the initial and final states, but also of the time of transition as well. When one-step transition probabilities are independent of the time variable (i.e., of the value of n), we say that the Markov process has *stationary transition probabilities.* We limit our discussion to such cases. So we assume that $P_{ij}^{n,n+1} = P_{ij}$ is independent of n and P_{ij} is the conditional probability that given it is in state i, it will move to state j in the next step.

Transition Probability Matrix

It is customary to arrange these numbers P_{ij} as a matrix,

$$P = \begin{bmatrix} P_{00} & P_{01} & \ldots & P_{0j} & \ldots \\ P_{10} & P_{11} & \ldots & P_{1j} & \ldots \\ \vdots & \vdots & & \vdots & \\ P_{i0} & P_{i1} & \ldots & P_{ij} & \ldots \\ \vdots & \vdots & \vdots & & \end{bmatrix}$$

and we refer to this matrix P as the *transition probability matrix* of the process.

The process is completely determined once this matrix P and the value or more generally the probability distribution of X_0 are specified.

Transition Probabilities

One of the main quantities we are interested in is the *n-step transition probabilities*

$$P_{ij}(n) = \Pr(X_n = j | X_0 = i).$$

These n-step transition probabilities satisfy the famous *Chapman–Kolmogorov equations,*

$$P_{ij}(n) = \sum_{k \in S} P_{ik}(m) P_{kj}(n-m), \quad m = 0, 1, 2, \ldots, n, \qquad (2.2)$$

for all $n \geq 0$ and $i, j \in S$. Let $P(n) \equiv (P_{ij}(n))$ be the *n-step probability matrix.* From (2.2), it follows that $P(n) = P^n$, $n = 0, 1, 2, \ldots$.

Periodicity of a Markov Chain

We define the period of state i, written $d(i)$, to be the greatest common divisor of all integers $n \geq 1$ for which $P_{ii}(n) = \Pr(X_n = i | X_0 = i) > 0$. If $P_{ii}(n) = 0$ for all $n \geq 1$ define $d(i) = 0$. A Markov chain in which each state has period one is called *aperiodic*. The vast majority of the Markov chains we deal with are aperiodic.

2.3 Random Walks

In discussing random walks it is an aid to intuition to speak about the state of the system as the position of a moving particle.

A one-dimensional random walk is a Markov chain in which the particle, if it is in state i, can in a single transition either stay in i or move to one of the adjacent states $i-1$, $i+1$. If the state space is taken as $S = \{-1, 0, 1, \ldots\}$, we suppose that the transition matrix of a random walk has the form

$$P = \begin{bmatrix} 1 & 0 & 0 & 0 & 0 & \ldots \\ q_0 & r_0 & p_0 & 0 & 0 & \ldots \\ 0 & q_1 & r_1 & p_1 & 0 & \ldots \\ \vdots & \vdots & \vdots & \vdots & \vdots & \end{bmatrix},$$

where $p_i, q_i > 0$ and $r_i \geq 0$ for $i \geq 1$, $p_0 > 0$, $q_0 \geq 0$, $r_0 \geq 0$, and $q_i + r_i + p_i = 1$, $i \geq 0$ and $q_{-1} = p_{-1} = 0$ and $r_{-1} = 1$, where

$$\begin{aligned} p_i &= \Pr(X_{n+1} = i+1 | X_n = i), \\ q_i &= \Pr(X_{n+1} = i-1 | X_n = i), \\ r_i &= \Pr(X_{n+1} = i | X_n = i). \end{aligned}$$

If $q_0 > 0$, we have a positive probability of entering state -1. This state is an absorbing state and once entered we will never leave it. If $q_0 = 0$ we ignore state -1 and we work with the matrix

$$P = \begin{bmatrix} r_0 & p_0 & 0 & 0 & \ldots \\ q_1 & r_1 & p_1 & 0 & \ldots \\ 0 & q_2 & r_2 & p_2 & \ldots \\ \vdots & \vdots & \vdots & \vdots & \end{bmatrix}.$$

The designation "random walk" seems apt since a realization of the process describes the path of a person suitably intoxicated moving randomly one step forward or backward.

Limiting Stationary Distribution

Suppose we have $q_0 = 0$ and we ignore state -1. Let us briefly discuss the behavior of $P_{ij}(n)$ as n becomes large. It can be shown that in this case the limits

$$\lim_{n\to\infty} P_{ij}(n) = \rho_j, \quad i, j \geq 0$$

exist and are independent of the initial state i. It turns out that the ρ_j are given by

$$\rho_j = \frac{\pi_j}{\sum_{k=0}^{\infty} \pi_k}, \quad j \geq 0,$$

where

$$\pi_0 = 1 \qquad \text{and} \qquad \pi_j = \frac{p_0 p_1 \cdots p_{j-1}}{q_1 q_2 \cdots q_j}, \quad j = 1, 2, \ldots.$$

In order that the sequence $\{\rho_j\}$ defines a distribution we must have $\sum_k \pi_k < \infty$ and then clearly $\sum_k \rho_k = 1$. We say that $\{\rho_j\}$ is the *limiting stationary distribution.* If $\sum_k \pi_k = \infty$ then all ρ_j are zero and we do not have a limiting stationary distribution.

2.4 Birth and Death Processes

Birth and death processes can be regarded as the continuous time analogues of random walks. They play a fundamental role in the theory and applications that embrace diverse fields, such as queueing and inventory models, chemical kinetics, and population dynamics.

Birth and Death Parameters

A *birth and death process* X_t is a Markov process with parameter set $\mathcal{T} = [0, \infty)$ on the state space $S = \{-1, 0, 1, 2 \ldots\}$ with stationary transition probabilities; i.e.,

$$P_{ij}(t) \equiv \Pr(X_{t+s} = j | X_s = i), \quad i, j \in S$$

is not dependent on s.

In addition we assume that the $P_{ij}(t)$ satisfy

1. $P_{i,i+1}(h) = \lambda_i h + o(h)$ as $h \downarrow 0, i \in S$;
2. $P_{i,i-1}(h) = \mu_i h + o(h)$ as $h \downarrow 0, i \geq 0$;
3. $P_{i,i}(h) = 1 - (\lambda_i + \mu_i)h + o(h)$ as $h \downarrow 0, i \in S$;
4. $P_{ij}(0) = \delta_{ij}$;
5. $P_{-1,-1}(t) = 1, P_{-1,i}(t) = 0, t \geq 0, i \neq -1,$
$\mu_0 \geq 0, \lambda_0 > 0, \lambda_i, \mu_i > 0, i \geq 1.$

The $o(h)$ may depend on i.

The parameters λ_i and μ_i are called, respectively, the *birth and death rates.* In Postulates 1 and 2 we are assuming that if the process starts in state i, then in a small interval of time the probabilities of going one state up or down are essentially proportional to the length of the interval. In Postulate 3 we are assuming other transitions are not allowed.

If $\mu_0 > 0$ then we have an *absorbing state* -1; once we enter -1 we can never leave it. If $\mu_0 = 0$ we have a *reflecting state* 0. After entering 0 we will always go back to state 1 after some time. In this case the state -1 can never be reached and so we ignore it and take $S = \{0, 1, 2, \ldots\}$.

Potential Coefficients

We suppose the following conditions on the parameters λ_i and μ_i,

$$C \equiv \sum_{n=0}^{\infty} \frac{1}{\lambda_n \pi_n} \sum_{i=0}^{n} \pi_i = \infty, \tag{2.3}$$

and

$$D \equiv \sum_{n=0}^{\infty} \frac{1}{\lambda_n \pi_n} \sum_{i=n+1}^{\infty} \pi_i = \infty, \tag{2.4}$$

where the *potential coefficients* π_n are given by

$$\pi_i = \frac{\lambda_0 \lambda_1 \cdots \lambda_{i-1}}{\mu_1 \mu_2 \cdots \mu_i}, \; i \geq 1$$

and $\pi_0 = 1$. In such a way we rule out the possibility of an explosion, i.e., reaching infinity in finite time. The quantity C can be interpreted as the mean of the first passage time from 0 to ∞ and D can be seen as the mean passage time from ∞ to state 0 [2]. Furthermore, in the following paragraphs we show that the transition probabilities will then satisfy some crucial differential equations. In most practical examples of birth and death processes these conditions are met and the birth and death process associated with the prescribed parameters is uniquely determined.

Infinitesimal Generator Matrix

All information of this process is put into a tridiagonal matrix

$$Q = \begin{bmatrix} -(\lambda_0 + \mu_0) & \lambda_0 & 0 & 0 & \cdots \\ \mu_1 & -(\lambda_1 + \mu_1) & \lambda_1 & 0 & \cdots \\ 0 & \mu_2 & -(\lambda_2 + \mu_2) & \lambda_2 & \cdots \\ \vdots & \vdots & \vdots & \vdots & \end{bmatrix}.$$

This matrix Q is called the *infinitesimal generator matrix* of the process. It is closely related to the infinitesimal generator $\mathcal{A}$ of the process. In the case of a birth and death process $\mathcal{A}$ is given by

$$\mathcal{A}f(i) = \lambda_i f(i+1) - (\lambda_i + \mu_i) f(i) + \mu_i f(i-1) \tag{2.5}$$

for all bounded real-valued functions $f \in \mathcal{B}(S)$. Note the close similarity with the difference equation of hypergeometric type (1.10).

Chapman–Kolmogorov Equations

Since the $P_{ij}(t)$ are probabilities we have $P_{ij}(t) \geq 0$ and

$$\sum_{j \in S} P_{ij}(t) = 1, \quad i \in S.$$

Using the Markovian property of the process we may also derive the *Chapman–Kolmogorov equations*

$$P_{ij}(t+s) = \sum_{k \in S} P_{ik}(t) P_{kj}(s), \quad i, j \geq 0.$$

This equation states that in order to move from state i to state j in time $t+s$, the process moves to some state $k \in S$ in time t and then from k to j in the remaining time s.

Initial Distribution

So far we have mentioned only the transition probabilities $P_{ij}(t)$. In order to obtain the probability that $X_t = n$ we must specify where the process starts or more generally the probability distribution for the initial state. We then have

$$\Pr(X_t = n) = \sum_{i \in S} q_i P_{in}(t),$$

where $q_i = \Pr(X_0 = i), i \in S$.

Waiting Times

With the aid of the above assumptions we may calculate the distribution of the random variable T_i, $i \geq 0$ which is the *waiting time* of X_t in state i; that is, given the process is in state i, what is the distribution of the time T_i until it first leaves state i. It turns out that T_i follows an exponential distribution $\text{Exp}(\lambda_i + \mu_i)$ with mean $(\lambda_i + \mu_i)^{-1}$.

According to Postulates 1 and 2, during a time duration of length h transition occurs from state i to $i+1$ with probability $\lambda_i h + o(h)$ and from state i to $i-1$ with probability $\mu_i h + o(h)$. It follows intuitively that, given that a transition occurs at time t, the probability that this transition is to state $i+1$ is $\lambda_i/(\lambda_i + \mu_i)$ and to state $i-1$ is $\mu_i/(\lambda_i + \mu_i)$. Thus the motion is analogous to that of a random walk except that transitions occur at random times rather than at fixed time periods.

Forward and Backward Equations

For $t \geq 0$ we define the matrix $P(t) = \{P_{ij}(t), i, j = 0, 1, 2, \ldots\}$, where as before

$$P_{ij}(t) = \Pr(X_t = j | X_0 = i), \ i, j \geq 0, t \geq 0.$$

This matrix will satisfy the so-called *backward differential equations*

$$P'(t) = Q \cdot P(t).$$

Because of the imposed condition (2.4) on D, $P(t)$ will also satisfy the *forward differential equations*

$$P'(t) = P(t) \cdot Q.$$

Limiting Stationary Distribution

Suppose we have $\mu_0 = 0$ and we ignore state -1. Let us briefly discuss the behavior of $P_{ij}(t)$ as t becomes large. It can be shown that in this case the limits

$$\lim_{t\to\infty} P_{ij}(t) = p_j$$

exist and are independent of the initial state i. It turns out that the p_j are given by

$$p_j = \frac{\pi_j}{\sum_{k=0}^{\infty} \pi_k}, \qquad j \geq 0.$$

In order that the sequence $\{p_j\}$ defines a distribution we must have $\sum_k \pi_k < \infty$ and then clearly $\sum_k p_k = 1$. We say that $\{p_j\}$ is the *limiting stationary distribution.* If $\sum_k \pi_k = \infty$ then all p_j are zero and we do not have a limiting stationary distribution.

2.5 Lévy Processes

Let $\{X_t, t \geq 0\}$ be a stochastic process and $0 \leq t_1 < t_2$. The random variable $X_{t_2} - X_{t_1}$ is called the *increment* of the process X_t over the interval $[t_1, t_2]$.

A stochastic process X_t is said to be a process with *independent increments* if the increments over nonoverlapping intervals (common endpoints are allowed) are stochastically independent. A process X_t is called a *stationary* or *homogeneous* process if the distribution of the increment $X_{t+s} - X_s$ depends only on t, but is independent of s. A stationary process with independent increments is called a *Lévy process.* From these assumptions it is also clear that a Lévy process satisfies the Markov property and thus a Lévy process is a special type of Markov process.

Infinitely Divisible Distribution

Let X_t be a Lévy process. We denote the characteristic function of the distribution of $X_{t+s} - X_s$ by

$$\phi(u, t) \equiv E[\exp(iu(X_{t+s} - X_s))].$$

It is known that $\phi(u,t)$ is *infinitely divisible* (i.e., for every positive integer n, it is the nth power of some characteristic function), and that

$$\phi(u,t) = [\phi(u,1)]^t.$$

We denote by

$$X_{t-} = \lim_{s \to t, s<t} X_s, \qquad t > 0,$$

the left limit process and by $\Delta X_t = X_t - X_{t-}$ the jump size at time t.

The following representation is valid for the characteristic function of an infinitely divisible distribution with finite variance.

Theorem 2 (Kolmogorov Canonical Representation) *The function $\phi(\theta)$ is the characteristic function of an infinitely divisible distribution with finite second moment if, and only if, it can be written in the form*

$$\psi(\theta) = \log \phi(\theta) = ic\theta + \int_{-\infty}^{+\infty} (e^{i\theta x} - 1 - i\theta x) \frac{dK(x)}{x^2}, \tag{2.6}$$

where c is a real constant and $K(y)$ is a nondecreasing and bounded function such that $K(-\infty) = 0$. The representation is unique.

For $x = 0$, the integrand $(e^{i\theta x} - 1 - i\theta x)/x^2$ is defined to be equal to $-(\theta^2/2)$. The function ψ is often called the *characteristic exponent* of the Lévy process.

A related important formula is the *Lévy–Khintchine formula* [16].

Theorem 3 (Lévy–Khintchine formula) *A function $\psi : R \to C$ is the characteristic exponent of an infinitely divisible distribution if and only if there are constants $a \in R$, $\sigma^2 \geq 0$, and a measure ν on $R \backslash \{0\}$ with $\int_{-\infty}^{+\infty} (1 \wedge x^2) \nu(dx) < \infty$ such that*

$$\psi(\theta) = ia\theta - \frac{\sigma^2}{2}\theta^2 + \int_{-\infty}^{+\infty} (\exp(i\theta x) - 1 - i\theta x 1_{\{|x|<1\}}) \nu(dx)$$

for every θ.

The measure ν is called the *Lévy measure.*

If we have an infinitely divisible distribution with characteristic function $\phi(\theta)$, we can define a Lévy process X_t through the relations

$$\exp(\psi_X(\theta)) \equiv \phi_X(\theta) \equiv E[\exp(i\theta X_1)] = \phi(\theta).$$

Note that if $\{Y_t, t \geq 0\}$ is a Lévy process with characteristic function

$$E\{e^{i\theta Y_t}\} = \exp\{t\psi_Y(\theta)\},$$

then also $X_t := At + B\,Y_{Ct}$ with $C > 0$ is a Lévy process determined by

$$\psi_X(\theta) = i\theta A + C\,\psi_Y(B\,\theta). \tag{2.7}$$

TABLE 2.1. Lévy Processes

Name	Distribution	$\phi(u,1)$	Notation
Brownian motion	$\mathrm{N}(0,t)$	$e^{-u^2/2}$	B_t
Poisson process	$\mathrm{P}(t)$	$\exp(e^{iu}-1)$	N_t
Gamma process	$\mathrm{G}(t,1)$	$(1-iu)^{-1}$	G_t
Pascal process	$\mathrm{Pa}(t,p)$	$(p/(1-(1-p)e^{iu}))$	P_t
Meixner process		$\left(\frac{\cos(a/2)}{\cosh((u-ia)/2)}\right)^{2\mu}$	H_t

In Table 2.1, we summarize the names we give to the process X_t according to the distribution of $X_{t+s} - X_s$.

Remark that the Pascal process is sometimes also called *a negative binomial process* and that the Poisson process in the literature is sometimes dependent on a parameter $\lambda > 0$, and in this case $\phi(u,1) = \exp(\lambda(e^{iu}-1))$. We always take $\lambda = 1$.

Brownian motion and the Poisson process are among the most studied stochastic processes. The Gamma process is used in insurance (see for example [31], [32], [39], [40], and [49]), in hydrology [85], and in finance [79], and has many other applications. The Meixner process was introduced in [103] and [101] and recently B. Grigelionis [52] derived an analogue of the Black–Scholes formula for it.

2.6 Diffusion Processes

Having considered birth and death processes in which all changes occur by jumps, we turn to the other extreme where the sample functions are (with probability one) continuous.

Diffusions

Diffusions are Markov processes in continuous time with state space $S = (a,b)$, $-\infty \leq a < b \leq +\infty$, having continuous sample paths. The prototype for diffusion processes is the standard Brownian motion (or Wiener process). This is the process with independent normally distributed increments. Its generator is given by

$$\mathcal{A}f(x) = \frac{1}{2}f''(x).$$

Its mean drift is equal to zero and the variance of its increments $B_{t+s} - B_s$ is equal to t.

Drift Coefficient and Diffusion Coefficient

One can imagine a Markov process X_t that has continuous sample paths but that is *not* a process with independent increments. Suppose that, given

$X_s = x$, for (infinitesimal) small times t, the displacement $X_{s+t} - X_s = X_{s+t} - x$ has mean and variance approximately $t\mu(x)$ and $t\sigma^2(x)$, respectively. Here $\mu(x)$ and $\sigma(x)$ are functions of the state x, and not constants as in the case of Brownian motion.

So we suppose that

$$\begin{aligned} E[X_{s+t} - X_s | X_s = x] &= t\mu(x) + o(t), \\ E[(X_{s+t} - X_s)^2 | X_s = x] &= t\sigma^2(x) + o(t), \\ E[|X_{s+t} - X_s|^3 | X_s = x] &= o(t) \end{aligned} \tag{2.8}$$

hold, as $t \to 0$, for every x in the state space S.

Note that these relations hold for Brownian motion with $\mu = 0$ and $\sigma^2 = 1$. A more general formulation of the existence of infinitesimal mean and variance parameters that does not require the existence of finite moments, is the following. For every $\epsilon > 0$ assume that

$$\begin{aligned} E[(X_{s+t} - X_s)1_{(|X_{s+t} - X_s| \leq \epsilon)} | X_s = x] &= t\mu(x) + o(t), \\ E[(X_{s+t} - X_s)^2 1_{(|X_{s+t} - X_s| \leq \epsilon)} | X_s = x] &= t\sigma^2(x) + o(t), \\ \Pr(|X_{s+t} - X_s| > \epsilon | X_s = x) &= o(t) \end{aligned} \tag{2.9}$$

hold as $t \to 0$ and where $1_{(A)} = 1$ if A holds and is zero otherwise. It is a simple exercise to show that (2.8) implies (2.9). However there are many Markov processes with continuous sample paths for which (2.9) holds, but not (2.8). For example, (2.8) does not hold for the diffusion $X_t = \exp(B_t^3)$, but (2.9) is satisfied.

We say that a Markov process X_t on the state space $S = (a, b)$ is a *diffusion with drift coefficient* $\mu(x)$ *and diffusion coefficient* $\sigma^2(x) > 0$, if it has continuous sample paths, and relations (2.9) hold for all $x \in S$.

Generator of a Diffusion Process

Next we look at the generator $\mathcal{A}$ of such a diffusion. A clear proof of the following proposition is given in [17].

Proposition 1 *Let X_t be a diffusion on $S = (a, b)$. Then, all twice continuously differentiable f, vanishing outside a closed bounded subinterval of S, belong to $\mathcal{D}_{\mathcal{A}}$, and for such f,*

$$\mathcal{A}f(x) = \mu(x)f'(x) + \frac{1}{2}\sigma^2(x)f''(x).$$

Note the close relation with the differential equation of hypergeometric type (1.6).

2.6.1 Calculation of Transition Probabilities

This section is a formal introduction to the calculations of the transition densities by *spectral methods*. Additional arguments have to be included to make things precise.

Consider an arbitrary diffusion on S with drift coefficient $\mu(x)$ and diffusion coefficient $\sigma^2(x)$. Define the function

$$\rho(x) = \frac{2c}{\sigma^2(x)} \exp\left(\int_{x_0}^{x} \frac{2\mu(z)}{\sigma^2(z)} dz\right), \quad x \in S,$$

where c is an arbitrary positive constant and x_0 is an arbitrarily chosen state. In fact, ρ is the solution of the differential equation (1.7),

$$(s(x)\rho(x))' = \tau(x)\rho(x),$$

with $s(x) = \sigma^2(x)/2$ and $\tau(x) = \mu(x)$.

In most cases, we choose c such that $\rho(x)$ is a probability density on S.

Consider the space $L^2(S, \rho)$ of real-valued functions on S that are square integrable with respect to the density $\rho(x)$. Let $f, g \in \mathcal{D}_{\mathcal{A}}$, zero outside a compact interval. Then, upon integration by parts, one obtains the following property for $\mathcal{A}$,

$$\langle \mathcal{A}f, g\rangle_\rho = \langle f, \mathcal{A}g\rangle_\rho,$$

where

$$\mathcal{A}f(x) = \frac{1}{2}\sigma^2(x)f''(x) + \mu(x)f'(x), \quad x \in S$$

and the inner product $\langle \cdot, \cdot \rangle_\rho$ is defined by

$$\langle f, g\rangle_\rho = \int_S f(x)g(x)\rho(x)dx. \tag{2.10}$$

Consider the case in which S is a closed and bounded interval. The idea behind this method is that if ϕ is an eigenfunction of $\mathcal{A}$ corresponding to an eigenvalue α, i.e.,

$$\mathcal{A}\phi = \alpha\phi, \tag{2.11}$$

then $u(t, x) = e^{\alpha t}\phi(x)$ solves the backward equation

$$\frac{\partial u}{\partial t} = e^{\alpha t}\alpha\phi(x) = e^{\alpha t}\mathcal{A}\phi(x) = \mathcal{A}u(t, x).$$

Likewise, if $u(t, x)$ is a superposition (linear combination) of such functions then the same will be true.

Suppose that the set of eigenvalues, counting multiplicities, is countable, say $\alpha_0, \alpha_1, \alpha_2, \ldots$ with corresponding eigenfunctions $\phi_0, \phi_1, \ldots$ of unit length; i.e., $\langle \phi_m, \phi_m\rangle_\rho^{1/2} = 1$. It is simple to check that eigenfunctions corresponding to distinct eigenvalues are orthogonal with respect to the inner product (2.10). Also if there is more than one linearly independent eigenfunction for a single eigenvalue, then these eigenfunctions can be orthogonalized by the Gram–Schmidt procedure. So $\phi_0, \phi_1, \ldots$ can be taken orthonormal.

If the set of finite linear combinations of eigenfunctions is complete in $L^2(S,\rho)$, then each $f \in L^2(S,\rho)$ has a Fourier expansion of the form

$$f = \sum_{n=0}^{\infty} \langle f, \phi_n \rangle_\rho \phi_n.$$

Consider the superposition defined by

$$u_f(t,x) = \sum_{n=0}^{\infty} e^{\alpha_n t} \langle f, \phi_n \rangle_\rho \phi_n(x).$$

Then u_f satisfies the backward equation

$$\frac{\partial}{\partial t} u_f(t,x) = \mathcal{A} u_f(t,x)$$

and the initial condition

$$u_f(0,x) = f(x).$$

However, the function

$$T_t f(x) = E[f(X_t)|X_0 = x] = \int_S f(y) p(t; x, dy)$$

also satisfies the same backward equation

$$\frac{\partial}{\partial t} T_t f(x) = \mathcal{A} T_t f(x).$$

and the same initial condition. So if there is *uniqueness* for a sufficiently large class of initial functions f then we get

$$T_t f(x) = u_f(t,x),$$

which means

$$\int_S f(y) p(t; x, dy) = \int_S f(y) \left(\sum_{n=0}^{\infty} e^{\alpha_n t} \phi_n(x) \phi_n(y) \right) \rho(y) dy.$$

In such cases, therefore, $p(t; x, dy)$ has a density $p(t; x, y)$ and it is given by

$$p(t; x, y) = \sum_{n=0}^{\infty} e^{\alpha_n t} \phi_n(x) \phi_n(y) \rho(y).$$

2.6.2 Examples

We highlight the spectral representation for a variety of diffusion processes. In fact it was proven by Mazet in [80] that the below-mentioned diffusions are in essence the only ones that can be associated with a family of orthogonal polynomials.

The Ornstein–Uhlenbeck Process

Suppose we have $\mu(x) = -x$ and $\sigma^2(x) = 2$; then we have

$$\mathcal{A}H_n(x/\sqrt{2}) = -nH_n(x/\sqrt{2}),$$

where $H_n(x)$ is the Hermite polynomial of degree n and the operator $\mathcal{A}$ is given by

$$\mathcal{A}f = f''(x) - xf'(x).$$

We then obtain the spectral representation for the transition density

$$p(t; x, y) = \frac{e^{-y^2/2}}{\sqrt{2\pi}} \sum_{n=0}^{\infty} e^{-nt} H_n(x/\sqrt{2}) H_n(y/\sqrt{2}) \frac{1}{2^n n!}. \tag{2.12}$$

The Laguerre Diffusion

Suppose we have $\mu(x) = -bx + c$ and $\sigma^2(x) = 2x$ where $0 < x < \infty$ and the constants satisfy $b, c > 0$. A straightforward but arduous calculation leads to

$$p(t; x, y) = \frac{b^c y^{c-1} e^{-by}}{\Gamma(c)} \sum_{n=0}^{\infty} e^{-nbt} L_n^{(c-1)}(bx) L_n^{(c-1)}(by) \frac{\Gamma(n+1)}{\Gamma(n+c)},$$

with $L_n(x)$ the Laguerre polynomial of degree n. Next we use the formula (see Theorem 5.1 in [110])

$$\sum_{n=0}^{\infty} \frac{\Gamma(n+1)}{\Gamma(n+\alpha+1)} L_n^{(\alpha)}(x) L_n^{(\alpha)}(y) w^n$$

$$= \frac{\exp\left(-\frac{(x+y)w}{1-w}\right)}{1-w} (-xyw)^{-\alpha/2} J_\alpha\left(\frac{2(-xyw)^{1/2}}{1-w}\right),$$

where J_α is the *Bessel function of the first kind of order* α defined by

$$J_\alpha(z) = \sum_{n=0}^{\infty} \frac{(-1)^n (z/2)^{\alpha+2n}}{n!\Gamma(n+\alpha+1)}.$$

And so we obtain

$$\begin{aligned} p(t; x, y) &= \frac{b^c y^{c-1} e^{-by}}{\Gamma(c)(1-e^{-bt})} \exp\left(\frac{e^{-bt}(-b)(x+y)}{1-e^{-bt}}\right) \\ &\quad \times (-b^2 e^{-bt} xy)^{-(c-1)/2} J_{c-1}\left(\frac{2b\sqrt{-xye^{-bt}}}{1-e^{-bt}}\right). \end{aligned}$$

The Jacobi Diffusion

Suppose we have $\mu(x) = \tau(x) = (\alpha - (\alpha + \beta)x)/2$ and $\sigma^2(x) = (1 - x)x$, where $0 < x < 1$ and $\alpha, \beta > 0$

In this case the spectral expansion is in terms of the Jacobi polynomials $P_n^{(\beta-1,\alpha-1)}(x)$,

$$p(t; x, y) = \frac{y^{\alpha-1}(1-y)^{\beta-1}}{B(\alpha, \beta)} \times \sum_{n=0}^{\infty} e^{-n(n+\alpha+\beta-1)t/2} P_n^{(\beta-1,\alpha-1)}(2x - 1) P_n^{(\beta-1,\alpha-1)}(2y - 1)\pi_n,$$

where

$$\pi_n = \frac{B(\alpha, \beta)(2n + \alpha + \beta - 1)n!\Gamma(n + \alpha + \beta - 1)}{\Gamma(n + \alpha)\Gamma(n + \beta)}$$

and

$$B(\alpha, \beta) = \frac{\Gamma(\alpha)\Gamma(\beta)}{\Gamma(\alpha + \beta)}$$

is the *Beta function.*

Notes

The Markov property was named after the Russian mathematician Andrey Andreyevich Markov (1856–1922). Lévy processes were named after the French probabilist Paul Lévy (1886–1971) and the first research goes back to the late 1920s with the study of infinitely divisible distributions. The history of Brownian motion goes back to the late 1820s. Robert Brown then observed that pollen particles in suspension were in constant irregular motion. In 1900, L. Bachlier considerd Brownian motion a possible model for stock market prices [7]. In 1905, it was considered by Albert Einstein as a model of particles in suspension. In 1923 Nobert Wiener defined and constructed Brownian motion rigorously for the first time. Therefore, Brownian motion is sometimes also called the Wiener process.

A standard reference work for an integrated introduction to measure theory and probability is [19]. A general introduction to stochastic processes can be found in [17], [64], and [65]. Continuous-time Markov processes and especially birth and death processes are treated in [2]. An up-to-date account of the theory of Lévy processes can be found in [16] and [98]. A thorough yet accessible reference work concerning the application of stochastics in finance and insurance is [97]. In [20], a comprehensive and self-contained treatment of the probabilistic theory behind the risk-neutral valuation principle and its applications in the theory of pricing and hedging of financial derivatives is given.

The Jacobi Diffusion

Suppose we have $\mu(x) = $ [illegible] (α) [illegible] $(\alpha+\beta)x$ [illegible] and $\sigma^2(x) = $ [illegible] where [illegible] and [illegible].

In this case the spectral expansion is in terms of the Jacobi polynomials $P_n^{(\alpha-1,\beta-1)}(x)$:

$$p(t;x,y) = \frac{y^{\alpha-1}(1-y)^{\beta-1}}{B(\alpha,\beta)} \times \sum_{n=0}^{\infty} [\text{illegible}]$$

where

$$[\text{illegible}] = \frac{[\text{illegible}]}{\Gamma(n+\alpha)\Gamma(n+\beta)}$$

and

$$B(\alpha,\beta) = \frac{\Gamma(\alpha)\Gamma(\beta)}{\Gamma(\alpha+\beta)}$$

is the Beta function.

Notes

The history of [illegible] stochastic processes [illegible] and [illegible]. Lévy processes were named after the French probabilist Paul Lévy (1886–1971) and the first research goes back to the late 1920s with the study of infinitely divisible distributions. The history of Brownian motion goes back to the year 1827. Robert Brown then observed that pollen particles in suspension were in constant irregular motion. In 1900, L. Bachelier considered Brownian motion as a possible model for stock-market prices [6]. In 1905, in a very famous paper, Albert Einstein [illegible] as a model of particles in suspension. In 1923 Norbert Wiener defined and constructed Brownian motion rigorously for the first time. Therefore Brownian motion is sometimes also called the Wiener process.

[illegible] reference works for an integrated introduction to [illegible] and [illegible] [illegible] [illegible] stochastic processes [illegible] [illegible] and [illegible]. Continuous-time Markov processes and especially birth and death processes are treated in [illegible]. An up-to-date account of the theory of Lévy processes can be found in [illegible] and [95]. A thorough yet accessible reference work concerning the application of stochastic calculus in finance and insurance is [illegible]. In [illegible] a comprehensive and self-contained treatment of the probabilistic theory behind the risk-neutral valuation principle and its applications in the theory of pricing and hedging of financial derivatives is given.

3

Birth and Death Processes, Random Walks, and Orthogonal Polynomials

The study of the time-dependent behavior of birth and death processes (BDP) involves many intricate and interesting orthogonal polynomials, such as Charlier, Meixner, Laguerre, Krawtchouk, and other polynomials from the Askey scheme; other famous orthogonal polynomials not in the Askey scheme could appear also, e.g., orthogonal polynomials related to the Roger–Ramanujan continued fraction [88]. In fact, the three-term recurrence relation lies at the heart of continued fractions, orthogonal polynomials, and birth and death processes. For birth and death processes with complicated birth and death rates, for example, when rates are state dependent or nonlinear, it is almost impossible to find closed form solutions of the transition functions. Due to the difficulties involved in analytical methods, it is pertinent to develop other techniques to gain insight into the behavior of the various system characteristics such as system size probabilities, expected system size, etc. The Karlin and McGregor representation [59], [74] of the transition probabilities, which uses a system of orthogonal polynomials satisfying a three-term recurrence relation involving the birth and death rates, is very useful in understanding the asymptotic behavior of the birth and death process. In fact these polynomials appear in some important distributions, namely, the (doubly) limiting conditional distributions. Also these polynomials play a fundamental role in the study of exponential ergodicity (see , e.g., [33] through [36]).

3.1 Karlin and McGregor Spectral Representation for Birth and Death Processes

Consider a birth and death process on the state space $S = \{-1, 0, 1, 2, \ldots\}$. The process is given by its birth and death rates λ_i, μ_i, $i \in S$ as described in Section 2.4. As before -1 is an absorbing state. We ignore this state if $\mu_0 = 0$ and 0 becomes in this case a reflecting state. Remember that we always impose Conditions (2.3) and (2.4).

Birth–Death Polynomials

In the analysis of birth and death processes, a prominent role is played by a sequence of polynomials $\{Q_n(x), n \geq 0\}$, called *birth–death polynomials.* They are determined uniquely by the recurrence relation

$$-xQ_n(x) = \mu_n Q_{n-1}(x) - (\lambda_n + \mu_n)Q_n(x) + \lambda_n Q_{n+1}(x), \quad n \geq 0, \quad (3.1)$$

together with $Q_{-1}(x) = 0$ and $Q_0(x) = 1$. Karlin and McGregor proved that the transition function P can be represented as

$$P_{ij}(t) = \Pr(X_t = j | X_0 = i) =$$
$$\pi_j \int_0^\infty e^{-xt} Q_i(x) Q_j(x) d\phi(x), \quad i, j = 0, 1, \ldots, t \geq 0, \quad (3.2)$$

where ϕ is a positive Borel measure with total mass 1 and with support on the nonnegative real axis; ϕ is called the *spectral measure* of P. Taking $t = 0$ in (3.2) one easily sees that the polynomials $\{Q_n(x), n \geq 0\}$ are orthogonal with respect to ϕ.

The representation (3.2) of $P_{ij}(t)$ is remarkable in that the dependence on t in the right-hand side is restricted entirely to the monotone decreasing exponential term e^{-tx} and that the dependence on i and j is factored as the product $Q_i(x)Q_j(x)$.

For certain choices of birth and death parameters, the equations in (3.1) are recognizable recursion relations defining classical orthogonal polynomials, for which the orthogonalizing measures are well known. Under Conditions (2.3) and (2.4), the spectral measure in (3.2) can only be this known orthogonalizing measure, and the birth and death transition function can be calculated directly from (3.2).

It is well known that Q_n has n positive, simple zeros $x_{n1} < x_{n2} < \ldots < x_{nn}$ and that the limits

$$\xi_i = \lim_{n \to \infty} x_{ni}, \quad i \geq 1$$

exist and satisfy $0 \leq \xi_i \leq \xi_{i+1} < \infty$. Furthermore, because of Conditions (2.3) and (2.4) the related moment problem is determinate and thus

$$\inf(\mathrm{supp}(\phi)) = \xi_1 \quad \text{and} \quad \phi(\{\xi_1\}) = \left(\sum_{k=0}^{\infty} \pi_k Q_k^2(\xi_1) \right)^{-1}, \quad (3.3)$$

where we denote by supp(ϕ) the support of the measure ϕ.

Some results follow directly from the spectral representation (3.2). For example, in the case $\mu_0 = 0$, we have $Q_i(0) = 1$, $i \geq 0$, and thus the limiting stationary distribution, if it exists, is equal to

$$p_j = \frac{\pi_j}{\sum_{k=0}^{\infty} \pi_k} = \lim_{t \to \infty} P_{ij}(t) = \pi_j Q_j(0) Q_i(0) \phi(\{0\}) = \pi_j \phi(\{0\}).$$

The following symmetry relation also follows directly from (3.2).

$$\pi_i P_{ij}(t) = \pi_j P_{ji}(t).$$

We say that $P_{ij}(t)$ is *weakly symmetric*.

Example: The $M/M/\infty$ Queue

The $M/M/\infty$ queue is a birth and death process with birth and death parameters given by

$$\lambda_n = \lambda, \qquad \mu_n = n\mu, \qquad n \geq 0,$$

where $\lambda, \mu > 0$ are the arrival and service rates, respectively. We have

$$\pi_j = \frac{(\lambda/\mu)^j}{j!}, \quad j \geq 0,$$

and the birth–death polynomials satisfy

$$0 = n\mu Q_{n-1}(x) + (x - \lambda - n\mu) Q_n(x) + \lambda Q_{n+1}(x), \quad n \geq 0,$$

where $Q_0(x) = 1$ and $Q_{-1}(x) = 0$. If we divide by μ, we obtain

$$0 = n Q_{n-1}(x) + \left(\frac{x}{\mu} - \frac{\lambda}{\mu} - n \right) Q_n(x) + \frac{\lambda}{\mu} Q_{n+1}(x), \quad n \geq 0.$$

Comparing this with the defining equations of the Charlier polynomials, $C_n(x; a)$,

$$0 = n C_{n-1}(x; a) + (x - a - n) C_n(x; a) + a C_{n+1}(x; a), \quad n \geq 0,$$

where $C_0(x; a) = 1$ and defining $C_{-1}(x; a) = 0$, we see that

$$Q_n(x) = C_n \left(\frac{x}{\mu}; \frac{\lambda}{\mu} \right), \quad n \geq 0.$$

The Charlier polynomials satisfy a generating function relation

$$\sum_{i=0}^{\infty} C_i(x; a) \frac{z^i}{i!} = e^z \left(1 - \frac{z}{a} \right)^x$$

and they are orthogonal with respect to the Poisson distribution $\mathrm{P}(a)$ supported on the nonnegative integers and given by

$$w_n = \frac{a^n}{n!}e^{-a}, \quad n = 0, 1, 2, \ldots. \tag{3.4}$$

But this means that the $Q_i(x)$, $i \geq 0$ are orthogonal with respect to the probability measure assigning the mass w_n given in (3.4) to the points $n\mu$, $n \geq 0$, so that

$$\begin{aligned} P_{ij}(t) &= \pi_j \sum_{n=0}^{\infty} e^{-t\mu n} Q_i(\mu n) Q_j(\mu n) w_n \\ &= \pi_j \sum_{n=0}^{\infty} e^{-t\mu n} C_i\left(n; \frac{\lambda}{\mu}\right) C_j\left(n; \frac{\lambda}{\mu}\right) \frac{(\lambda/\mu)^n}{n!} e^{-(\lambda/\mu)}. \end{aligned}$$

Using the generating function and the duality relation (1.16), we find

$$\begin{aligned} \sum_{j=0}^{\infty} P_{ij}(t) z^j &= \sum_{n=0}^{\infty} e^{-t\mu n} C_n(i; \lambda/\mu) \frac{(\lambda/\mu)^n}{n!} e^{-\lambda/\mu} \sum_{j=0}^{\infty} \pi_j C_j(n; \lambda/\mu) z^j \\ &= (1 - e^{\mu t} + z e^{-\mu t})^i \exp((\lambda/\mu)(1 - e^{-\mu t})(z - 1)). \end{aligned}$$

The stationary distribution is given by

$$p_i = \frac{\pi_i}{\sum_{k=0}^{\infty} \pi_k} = e^{-(\lambda/\mu)} \frac{(\lambda/\mu)^i}{i!},$$

the Poisson distribution $\mathrm{P}(\lambda/\mu)$.

This birth and death process with $\mu = 1$ plays an important role in Stein's theory for the Poisson distribution and is sometimes also called the immigration–death process.

Example: The Linear Birth and Death Process

We consider in this example birth and death processes with linear parameters [61] of the form

$$\lambda_n = (n + \beta)\lambda \quad \text{and} \quad \mu_n = n\mu, \quad \text{where } \lambda, \mu, \beta > 0. \tag{3.5}$$

A simple computation gives

$$\pi_j = \frac{(\beta)_j}{j!} \left(\frac{\lambda}{\mu}\right)^j, \quad j = 0, 1, 2, \ldots.$$

To proceed further, we have to consider three separate cases.

Case 1 ($\lambda < \mu$). After a few calculations, it can be seen that the birth–death polynomials are given in terms of the Meixner polynomials

$$Q_i(x) = M_i\left(\frac{x}{\mu - \lambda}; \beta, \frac{\lambda}{\mu}\right), \quad n = 0, 1, 2, \ldots.$$

But this means that the $Q_i(x), i \geq 0$ are orthogonal with respect to the probability distribution assigning mass

$$w_n = \left(1 - \frac{\lambda}{\mu}\right)^\beta \frac{(\beta)_n}{n!} \left(\frac{\lambda}{\mu}\right)^n, \quad n = 0, 1, 2, \ldots,$$

at the points $(\mu - \lambda)n$, $n = 0, 1, 2, \ldots$.
Case 2 ($\lambda > \mu$). In this case the birth–death polynomials are given again in terms of the Meixner polynomials

$$Q_i(x) = \left(\frac{\mu}{\lambda}\right)^i M_i\left(\frac{x}{\lambda - \mu} - \beta; \beta, \frac{\mu}{\lambda}\right), \quad i = 0, 1, 2, \ldots$$

and thus the $Q_i(x), i \geq 0$ are orthogonal with respect to the probability distribution assigning mass

$$w_n = \left(1 - \frac{\mu}{\lambda}\right)^\beta \frac{(\beta)_n}{n!} \left(\frac{\mu}{\lambda}\right)^n, \quad n = 0, 1, 2, \ldots,$$

at the points $(n + \beta)(\lambda - \mu)$, $n = 0, 1, 2, \ldots$.
Case 3 ($\lambda = \mu$). After a few calculations, it can be seen that the birth–death polynomials are given in terms of the Laguerre polynomials

$$Q_i(x) = \frac{i!}{(\beta)_i} L_i^{(\beta-1)}(x/\lambda), \quad n \geq 0.$$

Therefore the $Q_i(x), i \geq 0$ are orthogonal on $[0, \infty)$ with respect to the density function of the Gamma distribution $\mathrm{G}(\beta, \lambda)$.

$$f(x) = \frac{1}{\lambda^\beta \Gamma(\beta)} x^{\beta-1} e^{-x/\lambda}.$$

Example: A Quadratic Model

Next we look at the birth and death process X_t with quadratic birth and death rates

$$\lambda_n = (N - n)(a - n) \quad \text{and} \quad \mu_n = n(b - (N - n)), \quad n = 0, 1, \ldots, N,$$

respectively, with $a, b \geq N$.

It is not so difficult to see that this process has as its stationary distribution, the hypergeometric distribution $\mathrm{HypII}(a, b, N)$. Indeed the stationary distribution is given by

$$p_i \equiv \lim_{t \to \infty} \Pr(X_t = i | X_0 = k) = \frac{\pi_i}{\sum_{j=0}^N \pi_j} = \frac{\binom{a}{i}\binom{b}{N-i}}{\binom{a+b}{N}}, \quad i = 0, 1, \ldots, N.$$

Using Karlin and McGregor spectral representation (3.2), we can write the transition probabilities of this process X_t as follows.

$$\Pr(X_t = j | X_0 = i) = \pi_j \sum_{x=0}^N e^{\lambda(x)t} R_i(\lambda(x); a, b, N) R_j(\lambda(x); a, b, N) \rho_x,$$

where

$$\rho_x = \frac{\binom{N-b-1}{N} N!(-N)_x(-a)_x(2x-a-b-1)}{(-1)^x x!(-b)_x(x-a-b-1)_{N+1}},$$

and

$$\lambda(x) = x(x-a-b-1)$$

and the R_i are the dual Hahn polynomials defined by

$$\begin{aligned} R_i(\lambda(x)) &\equiv R_i(\lambda(x); a, b, N) \\ &= {}_3\tilde{F}_2(-i, -x, x-a-b-1; -a, -N; 1), \end{aligned}$$

for $i = 0, 1, \ldots, N$.

Example: The Ehrenfest Model

Karlin and McGregor [63] also derive the exact solution for the *Ehrenfest model*, which is a birth and death process with finite-state space $S = \{0, 1, 2, \ldots, N\}$ and parameters

$$\lambda_n = (N-n)p \quad \text{and} \quad \mu_n = nq, \quad 0 \le n \le N,$$

where $0 < p < 1$ and $q = 1 - p$. In this case the orthogonal system involved consists of the Krawtchouk polynomials

$$Q_n(x) = K_n(x; N, p), \quad 0 \le n \le N,$$

where x is a discrete variable ranging over the integers $x = 0, 1, \ldots, N$, and the spectral measure is the binomial distribution $\text{Bin}(N, p)$ placing mass $\binom{N}{x} p^x q^{N-x}$ at the points $x = 0, 1, \ldots, N$.

Other examples, with representations in terms of, for example, the Racah polynomials and other orthogonal polynomials not necessary in the Askey scheme, can be found in the literature; see, for example, [88] and [6].

3.2 Limiting Conditional Distributions for Birth and Death Processes

3.2.1 Absorbing Case

Limiting Conditional Distribution

If $\mu_0 > 0$ then -1 is an absorbing state. We can then consider the *limiting conditional distribution* $\lim_{t\to\infty} r_{ij}(t)$, with

$$r_{ij}(t) \equiv \Pr(X_t = j | X_0 = i, t < T < \infty), \tag{3.6}$$

where T denotes the time of absorption in state -1. This is the limiting probability of being in state j at time t given that the process is not absorbed at time t, but that absorption in state -1 occurs eventually.

Doubly Limiting Conditional Distribution

The *doubly limiting conditional distribution* $\lim_{t\to\infty}\lim_{s\to\infty} r_{ij}(t,s)$, with

$$r_{ij}(t,s) \equiv \Pr(X_t = j | X_0 = i, t+s < T < \infty) \tag{3.7}$$

and where T again denotes the time of absorption in state -1, is also of interest. This is the limiting probability of being in state j given that the process will not leave $\{0,1,2,3,\ldots\}$ in the distant future, but that absorption in state -1 occurs eventually.

Recalling the interpretation of ξ_1 in (3.3), we have the following result.

Theorem 4 *Suppose Conditions (2.3) and (2.4) hold. The limiting conditional distribution of (3.6) and the doubly limiting conditional distribution of (3.7) are given by*

$$\lim_{t\to\infty} r_{ij}(t) = \frac{\pi_j Q_j(\xi_1)}{\sum_{k=0}^{\infty}\pi_k Q_k(\xi_1)}, \quad \lim_{t\to\infty}\lim_{s\to\infty} r_{ij}(t,s) = \frac{\pi_j Q_j^2(\xi_1)}{\sum_{k=0}^{\infty}\pi_k Q_k^2(\xi_1)},$$

respectively, with the interpretation that the limit is 0 *whenever the sum in the denominator diverges.*

Proof: For the first limiting conditional distribution, we refer to [68].

For the second we have

$$\begin{aligned}
r_{ij}(t,s) &= \Pr(X_t = j | X_0 = i, t+s < T < \infty) \\
&= \frac{\Pr(X_t = j, t+s < T < \infty | X_0 = i)}{\Pr(t+s < T < \infty | X_0 = i)} \\
&= P_{ij}(t)\frac{\Pr(t+s < T < \infty | X_t = j)}{\Pr(t+s < T < \infty | X_0 = i)} \\
&= P_{ij}(t)\frac{\Pr(s < T < \infty | X_0 = j)}{\Pr(t+s < T < \infty | X_0 = i)} \\
&= P_{ij}(t)\frac{\int_0^\infty e^{-sx} Q_j(x) x^{-1} d\phi(x)}{\int_0^\infty e^{-(s+t)x} Q_i(x) x^{-1} d\phi(x)}.
\end{aligned}$$

The last equality follows by the representation presented in Equation (3.7) of [68]. Note that because $\int_0^\infty x^{-1} d\phi(x) < \infty$ (see [59]), both integrals in the last equation converge. Karlin and McGregor [60] show in the proof of their Theorem 11 that for any continuous function f,

$$\lim_{t\to\infty}\frac{\int_0^\infty e^{-xt} f(x) d\phi(x)}{\int_0^\infty e^{-xt} d\phi(x)} = f(\xi_1).$$

So we have,

$$\lim_{s\to\infty} r_{ij}(t,s) = P_{ij}(t)\frac{Q_j(\xi_1)}{e^{-\xi_1 t} Q_i(\xi_1)}$$

$$
\begin{aligned}
&= \frac{Q_j(\xi_1)}{Q_i(\xi_1)}\pi_j \int_0^\infty e^{-(x-\xi_1)t} Q_i(x)Q_j(x)d\phi(x) \\
&= \frac{Q_j(\xi_1)}{Q_i(\xi_1)}\pi_j [Q_i(\xi_1)Q_j(\xi_1)\phi(\{\xi_1\}) + \\
&\quad \int_{(\xi_1,\infty)} e^{-(x-\xi_1)t} Q_i(x)Q_j(x)d\phi(x)].
\end{aligned}
$$

This gives us using the Dominated Convergence Theorem

$$\lim_{t\to\infty}\lim_{s\to\infty} r_{ij}(t,s) = \phi(\{\xi_1\})\pi_j Q_j^2(\xi_1).$$

Using (3.3), the theorem follows. ◇

Example: The Linear Birth and Death Process

We illustrate these results by computing the (doubly) limiting conditional distribution for the linear birth and death process with rates,

$$\lambda_n = \lambda(n+1) \quad \text{and} \quad \mu_n = \mu(n+1), \qquad n = 0, 1, 2, \ldots.$$

First let $\lambda < \mu$. So absorption is certain, and we are computing

$$\lim_{t\to\infty} \Pr(X_t = j | X_0 = i, t < T)$$

and

$$\lim_{t\to\infty}\lim_{s\to\infty} \Pr(X_t = j | X_0 = i, t + s < T).$$

After some calculations one can see that the polynomials can be expressed in terms of Meixner polynomials and are orthogonal with respect to a measure which has point masses located at the points $(\mu - \lambda)(n+1)$, $n = 0, 1, 2, \ldots$. So we have $\xi_1 = \mu - \lambda$. It is easy to calculate by induction

$$\pi_n = \left(\frac{\lambda}{\mu}\right)^n \frac{1}{n+1} \quad \text{and} \quad Q_n(\xi_1) = (n+1).$$

By an easy calculation

$$\lim_{t\to\infty} r_{ij}(t) = \left(\frac{\lambda}{\mu}\right)^j \left(1 - \frac{\lambda}{\mu}\right)^{-1}$$

and

$$\lim_{t\to\infty}\lim_{s\to\infty} r_{ij}(t,s) = \left(1 - \frac{\lambda}{\mu}\right)^2 (j+1)\left(\frac{\lambda}{\mu}\right)^j.$$

We see that the limiting conditional distribution is the geometric distribution $\text{Geo}(\lambda/\mu)$ and that the doubly limiting conditional distribution has a Pascal distribution $\text{Pa}(2, \lambda/\mu)$.

A similar calculation gives us for $\lambda > \mu$,

$$\lim_{t\to\infty} r_{ij}(t) = \left(1 - \frac{\mu}{\lambda}\right)^{-1} \left(\frac{\mu}{\lambda}\right)^j$$

and

$$\lim_{t\to\infty} \lim_{s\to\infty} r_{ij}(t,s) = \left(1 - \frac{\mu}{\lambda}\right)^2 (j+1) \left(\frac{\mu}{\lambda}\right)^j.$$

Here we use that $\xi_1 = \lambda - \mu$ and $Q_n(\lambda - \mu) = (n+1)(\mu/\lambda)^n$.

3.2.2 Reflecting Case

Limiting Stationary Distribution

If $\mu_0 = 0$ then we ignore state -1 and state 0 is a reflecting state. The limiting stationary distribution of the process, if it exists ($\sum_k \pi_k < \infty$), is given by

$$p_j = \lim_{t\to\infty} \Pr(X_t = j | X_0 = i) = \frac{\pi_j}{\sum_{k=0}^{\infty} \pi_k} = \pi_j \phi(\{0\}).$$

So the stationary distribution exists whenever ϕ has positive mass in 0.

Limiting Conditional Distribution

If $\sum_k \pi_k = \infty$, we do not have a limiting stationary distribution. We then can consider the *limiting conditional distribution* $\lim_{t\to\infty} \bar{r}_{ij}(t)$, with

$$\bar{r}_{ij}(t) \equiv \Pr(X_t = j | X_0 = i, t < S), \tag{3.8}$$

and where S denotes the last exit time from state 0. This is the limiting probability of being in state j given that the process has not started the drift to infinity.

Doubly Limiting Conditional Distribution

The *doubly limiting conditional distribution* $\lim_{t\to\infty} \lim_{s\to\infty} \bar{r}_{ij}(t,s)$, with

$$\bar{r}_{ij}(t,s) \equiv \Pr(X_t = j | X_0 = i, t + s < S), \tag{3.9}$$

and where S denotes the last exit time from state 0, can also be of interest. This is the limiting probability of being in state j given that the process will not drift to infinity in the distant future.

The next function plays an important role in determining the limiting conditional distribution. We define

$$G_i(t) = \Pr(t < S | X_0 = i), \qquad i \geq 0, t \geq 0.$$

In [68] it is shown in Equation (4.7) that

$$G_i(t) = \frac{\int_0^{\infty} e^{-xt} x^{-1} Q_i(x) d\phi(x)}{\int_0^{\infty} x^{-1} d\phi(x)} < \infty,$$

provided the process is transient. We need the constants

$$a_i \equiv G_i(0) = \Pr(0 < S|X_0 = i), \qquad i = 0, 1, \ldots.$$

For the reflecting case we have

Theorem 5 *Suppose* $\sum_k \pi_k = \infty$*; (2.3) and (2.4) hold. The limiting conditional distribution of (3.8) and the doubly limiting conditional distribution of (3.9) are given by*

$$\lim_{t\to\infty} \bar{r}_{ij}(t) = \frac{a_j \pi_j Q_j(\xi_1)}{\sum_{k=0}^{\infty} a_k \pi_k Q_k(\xi_1)}$$

and

$$\lim_{t\to\infty} \lim_{s\to\infty} \bar{r}_{ij}(t, s) = \frac{\pi_j Q_j^2(\xi_1)}{\sum_{k=0}^{\infty} \pi_k Q_k^2(\xi_1)},$$

with the interpretation that the limit is 0 whenever the sum in the denominator diverges.

Proof: The proof for the first limiting conditional distribution can again be found in [68].

We are now going to calculate $\lim_{t\to\infty} \lim_{s\to\infty} \bar{r}_{ij}(t, s)$. In a similar fashion to that of the absorbing case we have

$$\begin{aligned} \bar{r}_{ij}(t, s) &= \Pr(X_t = j | X_0 = i, t + s < S) \\ &= P_{ij}(t) \frac{\Pr(s < S | X_0 = j)}{\Pr(t + s < S | X_0 = i)} \\ &= P_{ij}(t) \frac{G_j(s)}{G_i(t + s)} \\ &= P_{ij}(t) \frac{\int_0^\infty e^{-sx} Q_j(x) x^{-1} d\phi(x)}{\int_0^\infty e^{-(s+t)x} Q_i(x) x^{-1} d\phi(x)}. \end{aligned}$$

Hence by the same arguments as in the absorbing case, the theorem follows. ◇

Example: The Linear Birth and Death Process

Let us look at the linear birth and death process, with birth and death rates given by

$$\lambda_n = (n + 1)\lambda \quad \text{and} \quad \mu_n = \mu n, \qquad n \geq 0.$$

Suppose that $\lambda > \mu$, so that we have a drift to infinity. As seen before in (3.5), the birth–death polynomials for this process can be written down in the function of the Meixner polynomials, and the spectral measure has mass points located at $(n+1)(\lambda-\mu), n=0,1,\ldots$. So $\xi_1 = \lambda-\mu$. A simple computation leads to

$$\pi_n = \left(\frac{\lambda}{\mu}\right)^n \quad \text{and} \quad Q_n(\xi_1) = \left(\frac{\mu}{\lambda}\right)^n,$$

so one finds easily that

$$\frac{\pi_j Q_j^2(\xi_1)}{\sum_{n=0}^{\infty} \pi_n Q_n^2(\xi_1)} = \left(1-\frac{\mu}{\lambda}\right)\left(\frac{\mu}{\lambda}\right)^j, \qquad j=0,1,\ldots,$$

and hence the doubly limiting conditional distribution is the geometric distribution $\mathrm{Geo}(\mu/\lambda)$.

3.3 Karlin and McGregor Spectral Representation for Random Walks

Random Walk Polynomials

Suppose we have a random walk X_n with

$$P = \begin{bmatrix} r_0 & p_0 & 0 & 0 & 0 & \ldots \\ q_1 & r_1 & p_1 & 0 & 0 & \ldots \\ 0 & q_2 & r_2 & p_2 & 0 & \ldots \\ \vdots & \vdots & \vdots & \vdots & \vdots & \end{bmatrix}. \tag{3.10}$$

We assume that $p_j > 0$, $r_j \geq 0$ for $j \geq 0$ and $q_j > 0$ for $j \geq 1$. Moreover, we require $p_j + r_j + q_j = 1$ for $j \geq 1$, but we allow $q_0 \equiv 1 - p_0 - r_0 \geq 0$. If $q_0 = 0$ then P is stochastic (row sums equal to one) and the random walk X is said to have a *reflecting boundary* 0. If $q_0 > 0$ then the random walk has an ignored absorbing state that can be reached through state 0 only. The absorbing state, if present, is denoted by -1.

The random walk X_n is aperiodic if $r_j > 0$ for some j, and periodic with period 2 if $r_j = 0$ for all j.

We associate with X_n the polynomial sequence $\{Q_j(x), j = 0,1,\ldots\}$ defined by the recurrence relations

$$xQ_j(x) = q_j Q_{j-1}(x) + r_j Q_j(x) + p_j Q_{j+1}(x), \qquad j \geq 0, \tag{3.11}$$

together with $Q_0(x) = 1$ and $Q_{-1}(x) = 0$. The polynomials $Q_n(x)$ are called *random walk polynomials.* Recall the following important constants

$$\pi_0 = 1, \qquad \pi_j = \frac{p_0 p_1 \cdots p_{j-1}}{q_1 q_2 \cdots q_j}\ j \geq 1. \tag{3.12}$$

The polynomials $Q_j(x), j \geq 0$ are orthogonal with respect to a unique Borel measure ϕ on the interval $[-1,1]$, and the transition probabilities $P_{ij}(n) =$

$\Pr(X_{m+n} = j|X_m = j)$ can be represented in terms of the polynomials $Q_j(x)$ and the measure ϕ by

$$P_{ij}(n) = \pi_j \int_{-1}^{1} x^n Q_i(x) Q_j(x) d\phi(x); \tag{3.13}$$

see Karlin and McGregor [62].

If we let $n \to \infty$ we see that $P_{ij}(n)$ can only converge to a nonzero limit if $\phi(\{-1\}) = 0$ and $\phi(\{1\}) > 0$. We then have

$$P_{ij}(n) \to \pi_j Q_j(1) Q_i(1) \phi(\{1\}).$$

Using the fact that $Q_j(1) = 1$, $j \geq 1$ and $\phi(\{1\}) = 1/(\sum_{k=0}^{\infty} \pi_k Q_j^2(1))$, we see that

$$P_{ij}(n) \to \frac{\pi_j}{\sum_{k=0}^{\infty} \pi_k}.$$

3.4 Limiting Conditional Distributions for Random Walks

Limiting Conditional Distribution

Sometimes the chain is not positive recurrent but has an absorbing state. We can then consider the *limiting conditional distribution* $\lim_{n\to\infty} r_{ij}(n)$, with

$$r_{ij}(n) \equiv \Pr(X_n = j | X_0 = i, n < T < \infty), \tag{3.14}$$

where T denotes the time of absorption at state -1. This is the limiting probability of being in state j given that the process is at time n in $\{0, 1, 2, \ldots\}$, but that absorption in state -1 occurs eventually.

Doubly Limiting Conditional Distribution

The *doubly limiting conditional distribution* $\lim_{n\to\infty} \lim_{m\to\infty} r_{ij}(n, m)$, with

$$r_{ij}(n, m) \equiv \Pr(X_n = j | X_0 = i, n + m < T < \infty), \tag{3.15}$$

and where T again denotes the time of absorption at state -1, also can be of interest. This is the limiting probability of being in state j given that the process will not leave $\{0, 1, 2, \ldots\}$ in the distant future, but that absorption in state -1 occurs eventually.

We have the following theorem, where we let

$$\eta = \sup \operatorname{supp}(\phi)$$

and

$$C_k(\phi) = \frac{\int_{-1}^{0} (-x)^k d\phi(x)}{\int_0^1 x^k d\phi(x)}.$$

Theorem 6 *If X is aperiodic, then the limiting conditional distribution of (3.14) and the doubly limiting conditional distribution of (3.15) exist, if $\eta < 1$ and $C_k(\phi) \to 0$ as $k \to \infty$, in which case the limits are given by*

$$\lim_{n\to\infty} r_{ij}(n) = \frac{\pi_j Q_j(\eta)}{\sum_{k=0}^{\infty} \pi_k Q_k(\eta)} \text{ and } \lim_{n\to\infty}\lim_{m\to\infty} r_{ij}(n,m) = \frac{\pi_j Q_j^2(\eta)}{\sum_{k=0}^{\infty} \pi_k Q_k^2(\eta)}.$$

Proof: For the proof of the first limit we refer to [104].

For the second limit we proceed as in Theorems 3 and 4 of the continuous case. We have

$$\begin{aligned} r_{ij}(n,m) &= \Pr(X_n = j | X_0 = i, n+m < T < \infty) & (3.16)\\ &= P_{ij}(n)\frac{\Pr(m < T < \infty | X_0 = i)}{\Pr(n+m < T < \infty | X_0 = i)}. & (3.17)\end{aligned}$$

By Proposition 2 of [105] we have that if X is aperiodic then

$$\lim_{m\to\infty} \frac{\Pr(m < T < \infty | X_0 = j)}{\Pr(m+n < T < \infty | X_0 = i)} \tag{3.18}$$

exists if $\eta = 1$, or $\eta < 1$ and $C_k(\phi) \to 0$ as $k \to \infty$ and

$$\lim_{m\to\infty} \frac{\Pr(m < T < \infty | X_0 = j)}{\Pr(m+n < T < \infty | X_0 = i)} = \frac{Q_j(\eta)}{\eta^n Q_i(\eta)}. \tag{3.19}$$

So we have

$$\lim_{m\to\infty} r_{ij}(n,m) = P_{ij}(n)\frac{Q_j(\eta)}{Q_i(\eta)\eta^n}. \tag{3.20}$$

Thus by the representation theorem and because if X is aperiodic

$$\phi(\{-\eta\}) = 0 \tag{3.21}$$

(see, for example, [104], p. 28), we conclude

$$\begin{aligned} \lim_{n\to\infty}\lim_{m\to\infty} r_{ij}(n,m) &= \lim_{n\to\infty} \frac{Q_j(\eta)}{Q_i(\eta)}\pi_j \int_{-1}^{1} \left(\frac{x}{\eta}\right)^n Q_i(x)Q_j(x)d\phi(x)\\ &= Q_j^2(\eta)\pi_j\phi(\{\eta\})\\ &= \frac{\pi_j Q_j^2(\eta)}{\sum_{k=0}^{\infty} \pi_k Q_k^2(\eta)}.\end{aligned}$$

This completes the proof of the theorem. ◇

Notes

In the study of the doubly limiting conditional distribution of Chapter 3, we always imposed Conditions (2.3) and (2.4). In [68] one studies the limiting conditional distribution without always imposing (2.3) and (2.4). It may be possible to obtain similar results for the doubly limiting conditional distribution without imposing these conditions. Also in the study of the doubly limiting conditional distribution, we supposed in the absorbing case that absorption occurs eventually. In [68] results are stated for the limiting conditional distribution for cases where absorption does not necessarily occur. The determination of the doubly limiting conditional distribution, in the case where absorption is not certain, is still an open problem. In [6] some special cases of the Racah polynomials are used in the analysis of a finite birth and death process with quadratic rates. Although we could relate some rather artifical birth and death process to the general Racah polynomial, it still is a challenge to look for a more natural stochastic interpretation.

4
Sheffer Systems

Lévy processes appear in many areas, such as in models for queues, insurance risks, and more recently in mathematical finance. Historically, the first research goes back to the late 1920s with the study of infinitely divisible distributions. In mathematical finance an important role is played by martingales. They are, for example, used with the interpretation of a risk-neutral market. The famous option-pricing model of Black and Scholes uses special martingales, which are related to a generating function of Hermite polynomials. In this chapter we give a large class of new martingales based on Sheffer polynomials for some important Lévy processes.

4.1 Lévy–Sheffer Systems

Sheffer Polynomials

Using the classical *Faa di Bruno formula* [69] one can easily show that the equation

$$f(z)\exp(x\,u(z)) = \sum_{m=0}^{\infty} Q_m(x)\frac{z^m}{m!} \tag{4.1}$$

generates a family of polynomials $\{Q_m(x), m \geq 0\}$ when both functions $u(z)$ and $f(z)$ can be expanded in a formal power series and if $u(0) = 0$, $u'(0) \neq 0$, and $f(0) \neq 0$. The polynomials $Q_m(x)$ so defined are of exact degree m and are called *Sheffer polynomials.* Any set of such polynomials is called a *Sheffer set* since the first treatment of such polynomials was started by Sheffer [106] and [107].

Lévy–Sheffer Systems

Define τ as the inverse function of u, so that $\tau(u(z)) = z$. Then τ also can be expanded formally in a power series with $\tau(0) = 0$ and $\tau'(0) \neq 0$.

We now introduce an additional time parameter $t \geq 0$ into the polynomials defined in (4.1) by replacing the function $f(z)$ by $(f(z))^t$.

Definition 2 *A polynomial set $\{Q_m(x,t), m \geq 0, t \geq 0\}$ is called a* **Lévy–Sheffer system** *if it is defined by a generating function of the form*

$$\sum_{m=0}^{\infty} Q_m(x,t)\frac{z^m}{m!} = (f(z))^t \exp(xu(z)), \tag{4.2}$$

where

(i) $f(z)$ and $u(z)$ are analytic in a neighborhood of $z = 0$,

(ii) $u(0) = 0$, $f(0) = 1$, and $u'(0) \neq 0$, and

(iii) $1/f(\tau(i\theta))$ is an infinitely divisible characteristic function.

The quantity t can be considered to be a positive parameter; as such the function $Q_m(x,t)$ will also be a polynomial in t.

If Condition *(iii)* is satisfied, then there is a Lévy process $\{X_t, t \geq 0\}$ defined by the function

$$\phi(\theta) = \phi_X(\theta) = \frac{1}{f(\tau(i\theta))} \tag{4.3}$$

through the characteristic function as in Section 2.5. From the Kolmogorov representation in Theorem 2, the latter can be equivalently phrased in terms of the pair (c, K).

The basic link between the polynomials and the corresponding Lévy processes is the following *martingale equality*

$$E[Q_m(X_t,t) \mid X_s] = Q_m(X_s,s), \quad 0 \leq s \leq t, m \geq 0. \tag{4.4}$$

Indeed, taking generating functions, we find on the left-hand side of (4.4)

$$\begin{aligned}
&\sum_{m=0}^{\infty} E\{Q_m(X_t,t) \mid X_s\}\frac{z^m}{m!} \\
&\quad = E\left\{\sum_{m=0}^{\infty} Q_m(X_t,t)\frac{z^m}{m!} \mid X_s\right\} \\
&\quad = E\left\{(f(z))^t \exp(u(z)X_t) \mid X_s\right\} \\
&\quad = (f(z))^t \exp(u(z)X_s) E\left\{\exp(u(z)(X_t - X_s)) \mid X_s\right\}.
\end{aligned}$$

For the rightside of (4.4) we immediately find

$$\sum_{m=0}^{\infty} Q_m(X_s, s)\frac{z^m}{m!} = (f(z))^s \exp(u(z)X_s).$$

Combination of both expressions leads to the relationship

$$E\{\exp(u(z)(X_t - X_s)) \mid X_s\} = (f(z))^{s-t}.$$

If we compare this relationship with the equation determining the Lévy process

$$E\{\exp(i\theta(X_t - X_s)) \mid X_s\} = (\phi(\theta))^{t-s}.$$

then we realize that (4.4) will be satisfied if and only if (4.3) holds.

Example: The Laguerre Polynomials

The following generating function of a version of the Laguerre polynomials is well known [70].

$$\sum_{m=0}^{\infty} L_m^{(\alpha-m)}(y)w^m = (1+w)^\alpha \exp(-yw).$$

We identify the ingredients of this example.

$$\begin{cases} u(z) = -z, \\ f(z) = (1+z)^\alpha, \\ \phi(\theta) = (1-i\theta)^{-\alpha}. \end{cases}$$

The function $\phi(\theta)$ resembles the characteristic function of the infinitely divisible Gamma distribution. Let $\{G_t, t \geq 0\}$ be the Gamma process with $(G_1 = G)$,

$$E[\exp(i\theta G_t)] = \exp(t\psi_G(\theta)),$$

where $\psi_G(\theta) = \log\phi(\theta) = -\log(1-i\theta)$. Hence in (2.7) we take $A = 0$, $B = 1$, and $C = \alpha$ so that $X_t = G_{\alpha t}$. We derive the martingale property by putting $Q_m(x,t) = L_m^{(\alpha t - m)}(x)$ so that

$$E\left[L_m^{(\alpha t - m)}(G_{\alpha t}) \mid G_{\alpha s}\right] = L_m^{(\alpha s - m)}(G_{\alpha s}).$$

Example: The Actuarial Polynomials

The *Actuarial polynomials* are determined by the generating function [46]

$$\sum_{m=0}^{\infty} g_m^{(\lambda)}(y)\frac{w^m}{m!} = \exp(\lambda w + y(1-e^w)),$$

where $\lambda > 0$. We identify the ingredients of this example.

$$\begin{cases} u(z) = 1 - \exp(z), \\ f(z) = \exp(\lambda z), \\ \phi(\theta) = (1 - i\theta)^{-\lambda}. \end{cases}$$

Indeed, it easily follows that $\tau(z) = \log(1 - z)$. As before we can put $X_t = G_{\lambda t}$. With the identification $Q_m(x,t) = g_m^{(\lambda t)}(x)$ we arrive at the martingale property

$$E\left[g_m^{(\lambda t)}(G_{\lambda t}) \mid G_{\lambda s}\right] = g_m^{(\lambda t)}(G_{\lambda s}) \,.$$

This martingale relation seems to be new.

4.2 Sheffer Sets and Orthogonality

Meixner Set of Orthogonal Polynomials

If a set of polynomials is defined as in (4.1), some extra conditions have to be satisfied to make these polynomials orthogonal. In his historic paper, Meixner [81] determined all sets of orthogonal polynomials that also satisfy the generating function relation (4.1). As we need some of the ingredients of Meixner's approach, let us briefly sketch the construction.

Put $D = d/dx$ for the differential operator with respect to x. Relation (4.1) implies that

$$\tau(D)Q_m(x) = m\, Q_{m-1}(x)\,, \qquad m \geq 0.$$

This equation in turn leads to the relation

$$\tau(D)(xQ_m(x)) = \tau'(D)Q_m(x) + m\,x\,Q_{m-1}(x)\,, \qquad m \geq 0.$$

By Favard's theorem, the monic set $\{Q_m(x), m \geq 0\}$ will be orthogonal if and only if the polynomials satisfy a three-term recurrence relation

$$Q_{m+1}(x) = (x + l_{m+1})Q_m(x) + k_{m+1}Q_{m-1}(x), \tag{4.5}$$

where the numbers l_m are real and $k_m < 0, m \geq 2$. Apply $\tau(D)$ to (4.5). Subtract from this relation (4.5) for Q_m after multiplying it by m. We obtain

$$(1 - \tau'(D))Q_m(x) = \\ (l_{m+1} - l_m)mQ_{m-1}(x) + \left(\frac{k_{m+1}}{m} - \frac{k_m}{m-1}\right) m(m-1)Q_{m-2}(x).$$

If we shift m to $m+1$ in this equation and then again apply $\tau(D)$, we find

$$(1-\tau'(D))Q_m(x) = (l_{m+2}-l_{m+1})mQ_{m-1}(x) + \left(\frac{k_{m+2}}{m+1} - \frac{k_{m+1}}{m}\right) m(m-1)Q_{m-2}(x).$$

Comparing the last two formulas, we obtain the following relations

$$l_{m+1} - l_m = \lambda, \quad m \geq 1,$$

$$\frac{k_{m+1}}{m} - \frac{k_m}{m-1} = \kappa, \quad m \geq 2,$$

$$(1-\tau'(D))Q_m(x) = \lambda\tau(D)Q_m(x) + \kappa\tau^2(D)Q_m(x), \quad m \geq 0, \tag{4.6}$$

and (4.5) becomes

$$Q_{m+1}(x) = (x+l_1+m\lambda)Q_m(x)+m(k_2+(m-1)\kappa)Q_{m-1}(x), \quad m \geq 0, \tag{4.7}$$

where $k_2 < 0$ and $\kappa \leq 0$. From (4.6) follows

$$\tau'(y) = 1 - \lambda\tau(y) - \kappa\tau^2(y). \tag{4.8}$$

Furthermore we obtain from (4.7), using

$$f(z) = \sum_{m=0}^{\infty} Q_m(0)\frac{z^m}{m!},$$

the following relation for $f(z)$,

$$\frac{f'(z)}{f(z)} = \frac{k_2 z + l_1}{1-\lambda z - \kappa z^2}.$$

We define two quantities α and β by the equation

$$1 - \lambda z - \kappa z^2 = (1-\alpha z)(1-\beta z),$$

where $\alpha\beta \geq 0$. With these quantities we can rewrite the equation for f and obtain from (4.8) a differential equation for the function $u(z)$.

$$u'(z) = \frac{1}{(1-\alpha z)(1-\beta z)}, \qquad \frac{f'(z)}{f(z)} = \frac{\ell + kz}{(1-\alpha z)(1-\beta z)}, \tag{4.9}$$

where $\ell \in \mathbb{R}$ and $k \leq 0$ are constants. The solution of these equations is standard. We obtain

$$u(z) = \begin{cases} \frac{1}{\alpha-\beta}\log\left(\frac{1-\beta z}{1-\alpha z}\right), & \text{if } \alpha \neq \beta, \\ \frac{z}{1-\alpha z}, & \text{if } \alpha = \beta. \end{cases}$$

Note that the value for $\beta = \alpha$ is the limiting expression from the case $\beta \neq \alpha$ when $\beta \to \alpha$.

The explicit form of f is a bit more complicated.

$$\log f(z) = \begin{cases} -\frac{(k+\alpha\ell)\log(1-\alpha z)}{\alpha(\alpha-\beta)} + \frac{(k+\beta\ell)\log(1-\beta z)}{\beta(\alpha-\beta)}, & 0 \neq \alpha \neq \beta \neq 0, \\ \frac{k\log(1-\alpha z)}{\alpha^2} + \frac{k+\alpha\ell}{\alpha}\frac{z}{1-\alpha z}, & \alpha = \beta \neq 0, \\ -\frac{(k+\alpha\ell)\log(1-\alpha z)}{\alpha^2} - \frac{kz}{\alpha}, & \alpha \neq \beta = 0, \\ \frac{k}{2}z^2 + \ell z, & \alpha = \beta = 0. \end{cases}$$

Again, the last three forms are the obvious limiting cases of the first form. Also the function τ can be obtained explicitly, giving

$$\tau(v) = \begin{cases} \frac{\exp(\beta v)-\exp(\alpha v)}{\beta\exp(\beta v)-\alpha\exp(\alpha v)} & \text{if } \alpha \neq \beta, \\ \frac{v}{1+\alpha v} & \text{if } \alpha = \beta. \end{cases}$$

For each choice of the allowed pairs (α, β) we obtain a *Meixner set of orthogonal polynomials.* Their explicit expression turns up after we have introduced the martingale context.

Lévy–Meixner System

Let us apply the above to Equation (4.2). We call any resulting system a *Lévy–Meixner system.* Each one of these systems is therefore a Lévy–Sheffer system but with orthogonal polynomials. Since the explicit form of the functions f and τ is known, we can identify the ingredients in the Kolmogorov representation (2.6). This then automatically determines the underlying process.

Before embarking on the different subcases, we try to get as far as possible with the general quantities. The differential equation for ϕ follows from that for f. From $z = \tau(v)$ and $-\log f(z) = \log\phi(-iu(z))$ we easily find that

$$-\frac{f'(z)}{f(z)} = \frac{\phi'(-iu(z))}{\phi(-iu(z))}(-i)u'(z)$$

and hence by the equations (4.9),

$$\frac{\phi'(-iu(z))}{\phi(-iu(z))} = -i(\ell + kz).$$

Changing back to the argument θ we find

$$\frac{\phi'(\theta)}{\phi(\theta)} = -i(\ell + k\tau(i\theta)).$$

The last equation allows us to identify the quantities k and ℓ. Put $\theta = 0$; then $\phi'(0) = iE[X_1] = -i\ell$. A further derivation similarly yields that $k = -\text{Var}[X_1] < 0$. Henceforth, we write

$$\ell = -\mu, \qquad k = -\sigma^2 .$$

Of course we can also solve the differential equation for $\phi(\theta)$. We easily find

$$\log \phi(\theta) = i\mu\theta + \sigma^2 \int_0^{i\theta} \tau(z)\, dz.$$

The identification of c and K in the Kolmogorov representation is done as follows. By taking derivatives in (2.6) at $\theta = 0$ we see that $ic = \phi'(0) = iE[X_1] = -i\ell$, and hence $c = \mu$. Taking another derivative we get the equation

$$\int_{-\infty}^{\infty} \exp(i\theta x)\, dK(x) = \sigma^2 \tau'(i\theta),$$

which determines K uniquely. The result of the subsequent calculations are rather easy and lead to the explicit form

$$\int_{-\infty}^{\infty} \exp(i\theta x)\, dK(x) = \begin{cases} \frac{\sigma^2(\alpha-\beta)^2 \exp(i(\alpha+\beta)\theta)}{(\beta \exp(i\beta\theta) - \alpha \exp(i\alpha\theta))^2} & \text{if } \alpha \neq \beta, \\ \left(\frac{\sigma}{1+i\alpha\theta}\right)^2 & \text{if } \alpha = \beta. \end{cases} \tag{4.10}$$

We later verify that the function $\int_{-\infty}^{\infty} \exp(i\theta x)\, d(K(x)/K(\infty))$ is indeed a characteristic function for all possible values of α and β where $\alpha\beta \geq 0$.

To simplify the further analysis and without loss of generality, we make the following choice.

$$\ell = c = 0, \qquad k = -K(\infty) = -\sigma^2 = -1 . \tag{4.11}$$

From the identification we then get

$$\psi(\theta) = \log \phi(\theta) = \begin{cases} \frac{i\theta(\alpha+\beta) + \log((\alpha-\beta)/(\alpha e^{i\alpha\theta} - \beta e^{i\beta\theta}))}{\alpha\beta}, & \text{if } 0 \neq \alpha \neq \beta \neq 0, \\ \frac{i\theta}{\alpha} - \frac{\log(1+i\alpha\theta)}{\alpha^2}, & \text{if } \alpha = \beta \neq 0, \\ \frac{i\theta}{\alpha} - \frac{(1-\exp(-i\alpha\theta))}{\alpha^2}, & \text{if } \alpha \neq \beta = 0, \\ -\frac{\theta^2}{2}, & \text{if } \alpha = \beta = 0. \end{cases} \tag{4.12}$$

4.3 The Lévy–Meixner Systems

Our general approach now is to link all Meixner polynomials to a unique Lévy process. The departing form for the polynomials is (4.2), while for the Lévy process we take (2.6) and (4.12); i.e.,

$$\log E[e^{i\theta X_t}] = t\psi(\theta) = t\log\phi_X(\theta) = t\int_{-\infty}^{\infty}(e^{i\theta x} - 1 - i\theta x)\frac{dK(x)}{x^2},$$

where K is a probability measure. The two ingredients are linked by Equation (4.3), $\psi(\theta) = -\log f(\tau(i\theta))$.

The measure of orthogonality $\Psi_t(x)$, is also the distribution function of our Lévy process X_t. Indeed, by taking generating functions in

$$\int_{-\infty}^{\infty} Q_m(x,t)Q_n(x,t)d\Psi_t(x) = \delta_{mn}c_m(t)$$

and setting $n = 0$ we have

$$\int_{-\infty}^{\infty}(f(z))^t\exp(xu(z))d\Psi_t(x) = c_0 = 1.$$

Putting $u(z) = i\theta$ so that $z = \tau(i\theta)$ finally gives

$$\int_{-\infty}^{\infty}\exp(i\theta x)d\Psi_t(x) = \left(\frac{1}{f(\tau(i\theta))}\right)^t = E[\exp(i\theta X_t)].$$

We also use the *unit step distribution at the origin*

$$I(x) = \begin{cases} 0, & x < 0 \\ 1, & x \geq 0. \end{cases}$$

4.3.1 Brownian Motion–Hermite

This is the case where $\alpha = \beta = 0$ and it is by far the easiest. The fundamental quantities are

$$\begin{cases} u(z) = z, \\ f(z) = \exp(-z^2/2), \\ \psi(\theta) = -\theta^2/2. \end{cases}$$

The first two quantities give us

$$\sum_{m=0}^{\infty} Q_m(x;t)\frac{z^m}{m!} = \exp(zx - tz^2/2),$$

while the last quantity tells us that we are working with a Brownian motion $\{B_t, t \geq 0\}$ as the appropriate Lévy process. Also $K(x) = I(x)$ is easily derived.

The generating function of the Hermite polynomials, $H_m(y)$, is given by

$$\sum_{m=0}^{\infty} H_m(y)\frac{w^m}{m!} = \exp(2yw - w^2).$$

Identification of the two expressions requires

$$y = \frac{x}{\sqrt{2t}}, \qquad w = z\sqrt{\frac{t}{2}}.$$

We therefore find that for each $m \in \mathbb{N}$, $\left\{(t/2)^{m/2} H_m\left(B_t/\sqrt{2t}\right)\right\}$ is a martingale; i.e., for $0 \le s < t$,

$$E\left[H_m\left(\frac{B_t}{\sqrt{2t}}\right) \mid B_s\right] = \left(\frac{s}{t}\right)^{m/2} H_m\left(\frac{B_s}{\sqrt{2s}}\right).$$

The latter result is well known and can be found, e.g., in [53]. Actually, this relation has been the point of departure of Plucinska's work [89] where it was shown that the above relation characterizes the Hermite polynomial up to a change of scale.

4.3.2 Poisson Process–Charlier

This is the case $\alpha \neq \beta = 0$ and the fundamental quantities are

$$\begin{cases} u(z) = -\frac{1}{\alpha}\log(1-\alpha z), \\ f(z) = (1-\alpha z)^{1/\alpha^2}\exp(z/\alpha)\,, \\ \psi(\theta) = i\frac{\theta}{\alpha} - \frac{1}{\alpha^2}(1-\exp(-i\alpha\theta)). \end{cases}$$

The first two quantities give us

$$\sum_{m=0}^{\infty} Q_m(x;t)\frac{z^m}{m!} = \exp(zt/\alpha)(1-\alpha z)^{(t-\alpha x)/\alpha^2}\,.$$

Moreover (4.10) tells us that $\int_{-\infty}^{\infty}\exp(i\theta x)\,dK(x) = \exp(-i\alpha\theta)$ and hence that $K(x) = I(x+\alpha)$. From the form of $\psi(\theta)$ it seems natural to search for a connection with the Poisson process $\{N_t, t \ge 0\}$ for which $\psi_N(\theta) = \exp(i\theta) - 1$. From (2.7) we derive that

$$A = \alpha^{-1}, \qquad B = -\alpha, \qquad C = \alpha^{-2}$$

so that

$$X_t = \frac{t}{\alpha} - \alpha N_{t/\alpha^2}.$$

The Charlier polynomials are defined for $a > 0$ by their generating function

$$\sum_{m=0}^{\infty} C_m(y;a)\frac{w^m}{m!} = e^w\left(1-\frac{w}{a}\right)^y.$$

Identification of them with the polynomials in (4.2) requires

$$w = \frac{zt}{\alpha}, \qquad a = \frac{t}{\alpha^2}, \qquad y = \frac{t-\alpha x}{\alpha^2}$$

and hence

$$Q_m(x,t) = C_m\left(\frac{t-\alpha x}{\alpha^2}, \frac{t}{\alpha^2}\right)\left(\frac{t}{\alpha}\right)^m.$$

Replacing x by $X(t)$ we derive by (2.7) the martingale property

$$E\left[C_m\left(N_{t/\alpha^2}, \frac{t}{\alpha^2}\right) \mid N_{s/\alpha^2}\right] = \left(\frac{s}{t}\right)^m C_m\left(N_{s/\alpha^2}, \frac{s}{\alpha^2}\right).$$

This property of the Poisson process was already known for $m = 1, 2, 3$; (we refer to, e.g., [16] and [111]), and implicitly it can be found in the work of Ogura [87] and Engel [45].

4.3.3 Gamma Process–Laguerre

Now we look at the case where $\alpha = \beta \neq 0$. The primary quantities are

$$\begin{cases} u(z) = \frac{z}{1-\alpha z}, \\ f(z) = (1-\alpha z)^{-1/\alpha^2}\exp\left(-\frac{z}{\alpha(1-\alpha z)}\right), \\ \psi(\theta) = i\frac{\theta}{\alpha} - \frac{1}{\alpha^2}\log(1+i\alpha\theta). \end{cases}$$

The first two ingredients lead to the generating function of the polynomials

$$\sum_{m=0}^{\infty} Q_m(x;t)\frac{z^m}{m!} = \exp\left(\frac{z(\alpha x - t)}{\alpha(1-\alpha z)}\right)(1-\alpha z)^{-t/\alpha^2}.$$

Furthermore $\int_{-\infty}^{\infty}\exp(i\theta x)\,dK(x) = (1+i\alpha\theta)^{-2}$ easily yields

$$K(x) = \int_{-\infty}^{x/\alpha\wedge 0} |y|\exp(y)dy.$$

From the form of $\psi(\theta)$ we are led to a Gamma process $\{G_t, t \geq 0\}$ which is determined through the expression $\psi_G(\theta) = -\log(1-i\theta)$. The requested identification leads to the same expressions as for the Poisson case. Hence again

$$X_t = \frac{t}{\alpha} - \alpha G_{t/\alpha^2}.$$

The obvious set of orthogonal polynomials is now provided by the Laguerre polynomials defined for all a through the generating function

$$\sum_{m=0}^{\infty} L_m^{(a)}(y)w^m = (1-w)^{-a-1}\exp\left(\frac{yw}{w-1}\right).$$

The appropriate identification with the basic polynomials requires

$$w = \alpha z, \qquad a = \frac{t}{\alpha^2} - 1, \qquad y = \frac{t-\alpha x}{\alpha^2}$$

and hence

$$Q_m(x,t) = m!\alpha^m\, L_m^{(-1+t/\alpha^2)}\left(\frac{t-\alpha x}{\alpha^2}\right).$$

Replace x by X_t to derive the martingale property

$$E\left[L_m^{(-1+t/\alpha^2)}\left(G_{t/\alpha^2}\right) \mid G_{s/\alpha^2}\right] = L_m^{(-1+s/\alpha^2)}\left(G_{s/\alpha^2}\right).$$

4.3.4 Pascal Process–Meixner

Let us now look at the case where $0 \neq \alpha \neq \beta \neq 0, \alpha, \beta \in \mathbb{R}, \alpha\beta > 0$. This case needs a bit more algebra. We assume that $\alpha/\beta < 1$. The fundamental quantities are now

$$\begin{cases} u(z) = \frac{1}{\alpha-\beta}\log\left(\frac{1-\beta z}{1-\alpha z}\right), \\ f(z) = (1-\alpha z)^{1/(\alpha(\alpha-\beta))}(1-\beta z)^{-1/(\beta(\alpha-\beta))}, \\ \psi(\theta) = i\frac{\alpha+\beta}{\alpha\beta}\theta + \frac{1}{\alpha\beta}\log\left(\frac{\alpha-\beta}{\alpha\exp(i\alpha\theta)-\beta\exp(i\beta\theta)}\right). \end{cases}$$

For the basic polynomials we therefore obtain the expression

$$\sum_{m=0}^{\infty} Q_m(x;t)\frac{z^m}{m!} = (1-\alpha z)^{(t-\alpha x)/(\alpha(\alpha-\beta))}(1-\beta z)^{(\beta x-t)/(\beta(\alpha-\beta))}.$$

The form of $\int_{-\infty}^{\infty}\exp(i\theta x)\,dK(x)$ has been given in (4.10) and leads without too much difficulty to the expression

$$K(x) = \sum_{m=0}^{\infty}(m+1)\left(\frac{\alpha}{\beta}\right)^m\left(1-\frac{\alpha}{\beta}\right)^2 I(x+(m+1)(\beta-\alpha)).$$

The Lévy process that appears is a Pascal process $\{P_t, t \geq 0\}$ that is characterized by $\psi_P(\theta) = \log p(1-q\exp(i\theta))^{-1}$ with $0<p<1, p+q=1$. The identification requires a little calculation but leads to $q = \alpha/\beta$ and further to

$$A = \beta^{-1}, \qquad B = \alpha-\beta, \qquad C = (\alpha\beta)^{-1}$$

so that
$$X_t = \frac{t}{\beta} + (\alpha - \beta) P_{t/(\alpha\beta)}.$$

From the Meixner classification we can identify the polynomials with the Meixner polynomials defined for $b > 0$ and $0 < c < 1$ by the generating function
$$\sum_{m=0}^{\infty} M_m(y; b, c) \frac{(b)_m w^m}{m!} = \left(1 - \frac{w}{c}\right)^y (1 - w)^{-y-b}.$$

We find
$$w = \alpha z, \qquad b = \frac{t}{\alpha\beta}, \qquad c = \frac{\alpha}{\beta}, \qquad y = \frac{\beta x - t}{\beta(\alpha - \beta)}.$$

We therefore get
$$Q_m(x, t) = \left(\frac{t}{\alpha\beta}\right)_m \alpha^m M_m\left(\frac{\beta x - t}{\beta(\alpha - \beta)}; \frac{t}{\alpha\beta}, \frac{\alpha}{\beta}\right).$$

There results the martingale property for $0 \leq s < t$,
$$\begin{aligned} &E\left[M_m\left(P_{t/(\alpha\beta)}; \frac{t}{\alpha\beta}, \frac{\alpha}{\beta}\right) \mid P_{s/(\alpha\beta)}\right] \\ &\quad = \frac{(s/\alpha\beta)_m}{(t/\alpha\beta)_m} M_m\left(P_{s/(\alpha\beta)}; \frac{s}{\alpha\beta}, \frac{\alpha}{\beta}\right). \end{aligned} \tag{4.13}$$

4.3.5 Meixner Process–Meixner–Pollaczek

The final case is with $0 \neq \alpha, \beta = \overline{\alpha}$. It requires some delicate algebra. The fundamental relations are
$$\begin{cases} u(z) = \frac{1}{\alpha - \overline{\alpha}} \log\left(\frac{1 - \overline{\alpha} z}{1 - \alpha z}\right), \\ f(z) = (1 - \alpha z)^{1/(\alpha(\alpha - \overline{\alpha}))} (1 - \overline{\alpha} z)^{-1/(\overline{\alpha}(\alpha - \overline{\alpha}))}, \\ \psi(\theta) = i \frac{\alpha + \overline{\alpha}}{\alpha\overline{\alpha}} \theta + \frac{1}{\alpha\overline{\alpha}} \log\left(\frac{\alpha - \overline{\alpha}}{\alpha \exp(i\alpha\theta) - \overline{\alpha} \exp(i\overline{\alpha}\theta)}\right). \end{cases}$$

We obtain the following expression for the basic polynomials
$$\sum_{m=0}^{\infty} Q_m(x; t) \frac{z^m}{m!} = (1 - \alpha z)^{(t - \alpha x)/(\alpha(\alpha - \overline{\alpha}))} (1 - \overline{\alpha} z)^{(\overline{\alpha} x - t)/(\overline{\alpha}(\alpha - \overline{\alpha}))}.$$

The determination of the Lévy process is not as easy as in the previous cases. We recall, from Section 2.5, the Meixner process as the Lévy process $\{H_t, t \geq 0\}$, where $H_1 = H$ has the characteristic function
$$\phi_H(\eta) = \left(\frac{\cos(a/2)}{\cosh((\eta - ia)/2)}\right)^{2\mu},$$

where $a \in \mathbb{R}$ and $\mu > 0$. This characteristic function corresponds to a density of the form

$$f_H(x;\mu,a) = \frac{(2\cos(a/2))^{2\mu}}{2\pi\Gamma(2\mu)}\exp(ax)|\Gamma(\mu+ix)|^2, \quad x \in \mathbb{R},$$

as can be proved by relying on the integral expression [51]

$$\int_{-\infty}^{\infty}\Gamma(\gamma - it)\Gamma(\gamma + it)\exp(i\eta t)dt = 2\pi\Gamma(2\gamma)\exp(\gamma\eta)(1 + \exp(\eta))^{-2\gamma}, \tag{4.14}$$

where $\eta \in \mathbb{R}$ and $\gamma > 0$.

Since $\beta = \overline{\alpha}$ it is natural to write $\alpha = \rho\exp(i\zeta)$. We have to identify the function $\psi(\theta)$ above with a suitable variant of $\psi_H(\theta)$. To do that, we first rewrite $\psi(\theta)$ by using the expression for α. The argument within the logarithm in the expression for $\psi(\theta)$ is written in the form

$$\frac{\alpha - \overline{\alpha}}{\alpha\exp(i\alpha\theta) - \overline{\alpha}\exp(i\overline{\alpha}\theta)} = \exp(-i\theta\rho\cos\zeta)\frac{\sin\zeta}{\sin(\zeta + i\theta\rho\sin\zeta)}\,.$$

Hence, we put $a/2 = (\pi/2) - \zeta$ in the expression for $\psi_H(\eta)$. A tedious calculation then shows that we can take $\mu = 1$ and that

$$\psi(\theta) = i\frac{\theta}{\rho}\cos\zeta + \frac{1}{2\rho^2}\psi_H(2\rho(\sin\zeta)\theta)\,.$$

The identification between the processes $\{X_t, t \geq 0\}$ and $\{H_t, t \geq 0\}$ in (2.7) can then be achieved by the choice

$$A = \frac{1}{\rho}\cos\zeta, \qquad B = 2\rho\sin\zeta, \qquad C = (2\rho^2)^{-1}.$$

Hence we obtain

$$X_t = \frac{t}{\rho}(\cos\zeta) + 2\rho(\sin\zeta)H_{t/(2\rho^2)}.$$

The Meixner–Pollaczek polynomial is defined for $\lambda > 0$ and $0 < \zeta < \pi$ by

$$\sum_{m=0}^{\infty} P_m(y;\lambda,\zeta)\frac{w^m}{m!} = (1 - \exp(i\zeta)w)^{-\lambda+iy}\,(1 - \exp(-i\zeta)w)^{-\lambda-iy}\,.$$

Here the identification is simple and leads to

$$w = z\rho, \qquad \lambda = \frac{t}{2\rho^2}, \qquad y = \frac{x}{2\rho\sin\zeta} - \frac{t}{2\rho^2}\cot\zeta\,.$$

The equality

$$Q_m(x,t) = m!\rho^m P_m\left(\frac{x}{2\rho\sin\zeta} - \frac{t}{2\rho^2}\cot\zeta, \frac{t}{2\rho^2}, \zeta\right)$$

easily yields the martingale expression

$$E\left[P_m\left(H_{t/(2\rho^2)}; \frac{t}{2\rho^2}, \zeta\right) \mid H_{s/(2\rho^2)}\right] = P_m\left(H_{s/(2\rho^2)}; \frac{s}{2\rho^2}, \zeta\right).$$

It remains to determine K. Using the form $\alpha = \rho \exp(i\zeta)$ as above we get the form

$$\int_{-\infty}^{\infty} \exp(i\theta x)\, dK(x) = \left(\frac{\sin \zeta}{\sin(\zeta + i\theta\rho \sin \zeta)}\right)^2.$$

We can again use the integral in (4.14) for $\gamma = 1$. Replace first $\sin \zeta = \cos((\pi/2) - \zeta) =: \cos(a/2)$ and rewrite the denominator in the form

$$\sin(\zeta + i\theta\rho \sin \zeta) = \cosh\left(\frac{-2\rho\theta \cos(a/2) - ia}{2}\right)$$

to get an expression as in $\phi_H(\theta)$ above. A little algebra reveals that K has a derivative with expression

$$\frac{dK(y)}{dy} = \frac{\sin \zeta}{\pi\rho}\left|\Gamma\left(1 - \frac{iy}{2\rho \sin \zeta}\right)\right|^2 \exp\left(-\frac{y(\pi - 2\zeta)}{2\rho \sin \zeta}\right).$$

4.4 i.i.d. Sheffer Systems

There is a simpler approach possible when trying to obtain *discrete time martingales* from a Sheffer set. Let $X_1, X_2, \ldots$ be a sequence of i.i.d. variables with characteristic function $\phi(\theta)$. Put

$$S_n = X_1 + X_2 + \ldots + X_n$$

for the sum. The latter can be taken as a discrete analogue of a continuous-time Lévy process and it has a characteristic function given by

$$E[\exp(i\theta S_n)] = (\phi(\theta))^n.$$

Definition 3 *A polynomial set* $\{Q_m(x,n), m \geq 0, n \in \{0,1,2,\ldots\}\}$ *is called an* **i.i.d. Sheffer system** *if it is defined by a generating function of the form*

$$\sum_{m=0}^{\infty} Q_m(x,n)\frac{z^m}{m!} = (f(z))^n exp(xu(z)) \tag{4.15}$$

where

(i) $f(z)$ *and* $u(z)$ *are analytic in a neighborhood of* $z = 0$,

(ii) $u(0) = 0$, $f(0) = 1$, *and* $u'(0) \neq 0$, *and*

(iii) $1/f(\tau(i\theta))$ *is a characteristic function.*

The quantity n can be considered to be a discrete positive parameter; as such the function $Q_m(x, n)$ will also be a polynomial in n.

If Condition *(iii)* is satisfied, then there is a sequence of i.i.d. random variables $X_1, X_2, \ldots$ defined by the function

$$\phi(\theta) = \phi_{X_i}(\theta) = \frac{1}{f(\tau(i\theta))} \tag{4.16}$$

through the characteristic function. The basic link between the polynomials and the corresponding i.i.d. variables is the following *martingale equality*

$$E[Q_m(S_n, n) \mid S_k] = Q_m(S_k, k), \quad 0 \le k \le n, m \ge 0, \tag{4.17}$$

which can be proved completely analogous as in the Lévy–Sheffer case of Section 4.1.

4.4.1 Examples

Examples for this procedure can be found easily; let us give somewhat new unusual applications.

Example: Generalized Bernoulli Polynomials

The *generalized Bernoulli polynomials* are determined by the generating function [30]

$$\sum_{m=0}^{\infty} B_m^{(n)}(y) \frac{w^m}{m!} = \left(\frac{w}{e^w - 1} \right)^n e^{wy},$$

where $n \in \{0, 1, 2, \ldots\}$. We identify the ingredients of (4.15) and (4.16) for this example:

$$u(z) = z, \quad f(z) = \frac{z}{e^z - 1}, \quad \phi(\theta) = \frac{e^{i\theta} - 1}{i\theta}.$$

Hence we conclude that the generalized Bernoulli polynomials induce martingales for the sequence S_n where the X_i have a uniform distribution U(0, 1):

$$E[B_m^{(n)}(S_n) \mid S_k] = B_m^{(k)}(S_k), \qquad 0 \le k \le n.$$

Example: Krawtchouk Polynomials

The *monic Krawtchouk polynomials* are determined by the generating function [70]

$$\sum_{m=0}^{\infty} \tilde{K}_m(x; n, p) \frac{w^m}{m!} = (1 + qw)^x (1 - pw)^{n-x},$$

where $n \in \{0, 1, 2, \ldots\}$, $0 < p < 1$, and $p + q = 1$.

We identify as before to find

$$u(z) = \log\left(\frac{1+qz}{1-pz}\right), \quad f(z) = 1 - pz, \quad \phi(\theta) = pe^{i\theta} + (1-p).$$

Hence we conclude that the polynomials

$$Q_m(x,n) = \tilde{K}_m(x;n,p)$$

induce martingales for the sequence S_n with the X_i Bernoulli distributed with parameter $0 < p < 1$:

$$E\left[\tilde{K}_m(S_n;n,p) \mid S_k\right] = \tilde{K}_m(S_k;k,p), \qquad 0 \le k \le n.$$

Example: Euler Polynomials

The *Euler polynomials of order* λ are determined by the generating function ([46], p. 253)

$$\sum_{m=0}^{\infty} E_m^{(\lambda)}(x)\frac{w^m}{m!} = \left(\frac{2}{1+e^w}\right)^{\lambda} e^{wx},$$

where $\lambda \in \{0, 1, 2, \ldots\}$.

We now identify as before to find

$$u(z) = z, \quad f(z) = \left(\frac{2}{1+e^z}\right)^{\lambda}, \quad \phi(\theta) = \left(\frac{1+e^{i\theta}}{2}\right)^{\lambda}.$$

Hence we conclude that the polynomials $Q_m(x,n) = E_m^{(\lambda n)}(x)$ induce martingales for the sequence S_n where the X_i have a binomial distribution $\mathrm{Bin}(\lambda, 1/2)$:

$$E[E_m^{(n\lambda)}(S_n) \mid S_k] = E_m^{(k\lambda)}(S_k).$$

Example: Narumi Polynomials

Next we consider the polynomials with generating function

$$\sum_{m=0}^{\infty} N_m^{(a)}(x)w^m = \left(\frac{\log(1+w)}{w}\right)^{a} (1+w)^x,$$

where $a \in \{0, 1, 2, \ldots\}$. Except for a shift in index, we have a set of polynomials considered by Narumi ([46], p. 258).

We now identify as before to find

$$u(z) = \log(1+z), \quad f(z) = \left(\frac{\log(1+z)}{z}\right)^{a}, \quad \phi(\theta) = \left(\frac{e^{i\theta}-1}{i\theta}\right)^{a}.$$

Hence we conclude that the polynomials $Q_m(x,n) = m!N_m^{(an)}(x)$ form martingales for the sequence S_n with the X_i distributed as a sum of a independent random variables with a uniform distribution $\mathrm{U}(0,1)$:

$$E[N_m^{(an)}(S_n) \mid S_k] = N_m^{(ak)}(S_k).$$

4.5 Convolution Relations

As a consequence of these martingale relations we can obtain some nice formulas involving these polynomials. They are essentially convolution relations. Recall that we denote by $\tilde{Q}_n(x)$ the monic version of $Q_n(x)$.

Corollary 1 *We have for* $0 < s \leq t$,

$$
\begin{aligned}
\tilde{H}_m(x,s) &= \int_{-\infty}^{+\infty} \tilde{H}_m(x+y,t) \frac{\exp\left(-\frac{y^2}{2(t-s)}\right)}{\sqrt{2\pi(t-s)}} dy, \\
\tilde{C}_m(x;s) &= \sum_{i=0}^{\infty} \tilde{C}_m(x+i;t) e^{-t+s} \frac{(t-s)^i}{i!}, \\
\tilde{L}_m^{(s-1)}(x) &= \int_0^{\infty} \tilde{L}_m^{(t-1)}(x+y) \frac{1}{\Gamma(t-s)} y^{t-s-1} e^{-y} dy, \\
L_m^{(s-m)}(x) &= \int_0^{\infty} L_m^{(t-m)}(x+y) \frac{1}{\Gamma(t-s)} y^{t-s-1} e^{-y} dy, \\
\tilde{M}_m(x;s,q) &= \sum_{i=0}^{\infty} \tilde{M}_m(x+i;t,q) \binom{i+t-s-1}{i} \times \\
&\quad (1-q)^{t-s} q^i, \\
\tilde{P}_m(x;s,\zeta) &= \int_{-\infty}^{+\infty} \tilde{P}_m(x+y;t,\zeta) \frac{(2\sin\zeta)^{2(t-s)}}{2\pi\Gamma(2(t-s))} \times \\
&\quad e^{(2\zeta-\pi)y} |\Gamma(t-s+i\zeta y)|^2 dy, \\
g_m^{(s)}(x) &= \int_0^{\infty} g_m^{(t)}(x+y) \frac{1}{\Gamma(t-s)} y^{t-s-1} e^{-y} dy, \\
B_m^{(n-1)}(x) &= \int_0^1 B_m^{(n)}(x+y) dy, \\
\tilde{K}_m(x;n-1,p) &= q\tilde{K}_m(x;n,p) + np\tilde{K}_m(x+1;n,p), \\
2E_m^{(n-1)}(x) &= E_m^{(n)}(x) + E_m^{(n)}(x+1), \\
N_m^{(n-1)}(x) &= \int_0^1 N_m^{(n)}(x+y) dy.
\end{aligned}
$$

Proof: We only give the proof for the Lévy–Sheffer polynomials. The i.i.d. case is completely analogous. We start from the martingale relation (4.4). Given that $X(s)$ has the value x, we have

$$Q_m(x,s) = E[Q_m(X(t),t) \mid X(s) = x], \quad 0 \leq s \leq t, m \geq 0. \tag{4.18}$$

Note that by the homogeneity, $F_X(y;t-s) \equiv \Pr(X(t) - X(s) \leq y)$ only depends on $t-s$. Because of the independence of the increments, and the definition of expected values, we have for the right-hand side

$$E[Q_m(X(t),t) \mid X(s) = x] = E[Q_m(X(t) - X(s) + x;t) \mid X(s) = x]$$

$$
\begin{aligned}
&= E[Q_m(X(t) - X(s) + x; t)] \\
&= \int_S Q_m(x + y; t) dF_X(y; t - s),
\end{aligned}
$$

with S the support of the distribution of $X(t) - X(s)$. Filling in the appropriate distribution for each Lévy–Sheffer or i.i.d. Sheffer system gives the desired formulas. ◇

Notes

Lévy processes are used, for example, in insurance and mathematical finance. Martingales play a key role in these areas and are related to the notion of a fair game or a risk-neutral market. The Meixner–Pollaczek martingales are related to the Meixner process, which was recently used in the study of option price models [52]. Exploiting the well-known properties of the polynomials involved in financial models looks very promising. The Meixner process, which was introduced in [103] and [101], is one of the more complex processes and has many properties that reflect, for example, the behavior of stock prices, asset returns, or other financial objects. The appropriateness of the Meixner process, other processes such as the Gamma and Pascal, and combinations of such processes in concrete situations has to be checked. Such a study of appropriateness is given in [43] for hyberbolic motion.

Excellent accounts of martingales are found in [83] and [113]. Martingales were first studied by Paul Lévy in 1934. The word *martingale* has different meanings: 1. an article of harness, to control a horse's head (in Dutch also called "Teugels"); 2. a rope for guying down the jib boom to the dolphin striker; 3. a system of gambling: the "doubling-up" system, in which bets are doubled progressively.

Pommeret [91] considers multidimensional Lévy–Sheffer systems and relates them to multidimensional natural exponential families.

5

Orthogonal Polynomials in Stochastic Integration Theory

In this chapter we study orthogonal polynomials in the theory of stochastic integration. Some orthogonal polynomials in stochastic theory will play the role of ordinary monomials in deterministic theory. A consequence is that related polynomial transformations of stochastic processes involved will have very simple chaotic representations. In a different context, the orthogonalization of martingales, the coefficients of some other orthogonal polynomials will play an important role. We start with a reference to deterministic integration and then search for stochastic counterparts. We look at integration with respect to Brownian motion, the compensated Poisson process, and the binomial process. Next we develop a chaotic and predictable representation theory for general Lévy processes satisfying some weak condition on its Lévy measure. It is in this representation theory that we need the concept of strongly orthogonal martingales and orthogonal polynomials come into play. Examples include the Gamma process, where again the Laguerre polynomials turn up, the Pascal process, with again the Meixner polynomials, and the Meixner process, with as expected the Meixner–Pollaczek polynomials. When we look at combinations of pure-jump Lévy processes and Brownian motion, an inner product with some additional weight in zero plays a major role. We give an example with the Laguerre-type polynomials introduced by L. Littlejohn.

5.1 Introduction

Deterministic Integration

In classical deterministic integration theory the polynomials

$$p_n(x) = \frac{x^n}{n!}, \quad n \geq 0$$

play a special role because they satisfy

$$\int_0^t p_n(x)dx = p_{n+1}(t).$$

Iterating this relation gives

$$\int_0^t \int_0^{t_n} \cdots \int_0^{t_1} dt_0 \dots dt_{n-1}dt_n = p_{n+1}(t).$$

Also the exponential function $\exp(x)$ has a special role because of the equation

$$\int_0^t \exp(x)dx = \exp(t) - \exp(0) = \exp(t) - 1.$$

Note too the relation between $\exp(x)$ and the $p_n(x)$,

$$\exp(x) = \sum_{n=0}^{\infty} p_n(x).$$

We are looking for analogues for stochastic theory. A good introduction to stochastic integration with respect to a broad class of processes (semimartingales) is [94]. We refer to this book for all definitions of stochastic integrals that are not explicitly stated.

It turns out that the role of p_n is taken by some orthogonal polynomials related to the distribution of the integrator.

Integration with Respect to Brownian Motion

The most studied stochastic case is integration with respect to Brownian motion $\{B_t, t \geq 0\}$, where B_t has a normal distribution $\mathrm{N}(0, t)$. The notion of multiple stochastic integration for this process was first introduced by Norbert Wiener. It is well known that in Ito integration theory with respect to standard Brownian motion, the Hermite polynomials play the role of the p_n [56]. We have

Theorem 7

$$\int_0^t \tilde{H}_n(B_s; s)dB_s = \frac{\tilde{H}_{n+1}(B_t; t)}{n+1}, \tag{5.1}$$

where $\tilde{H}_n(x;t) = (t/2)^{n/2} H_n(x/\sqrt{2t})$ *is the monic Hermite polynomial with parameter* t.

Note that the monic Hermite polynomials $\tilde{H}_n(x,t)$ are orthogonal with respect to the normal distribution $N(0,t)$, the distribution of B_t. Note also that because stochastic integrals are martingales, we recover the results of Section 4.3.1; i.e., $\{\tilde{H}_n(B_t;t)\}$ are martingales.

Geometric Brownian Motion

The generating function of the monic Hermite polynomials $\tilde{H}_n(x;t)$ is given by

$$\sum_{n=0}^{\infty} \tilde{H}_n(x;t)\frac{z^n}{n!} = \exp(-tz^2/2 + zx). \tag{5.2}$$

Using this generating function (5.2), one can easily see that the role of the exponential function is now taken by the function

$$Y(B_t,t) = \sum_{n=0}^{\infty} \frac{\tilde{H}_n(B_t;t)}{n!} = \exp(-t/2 + B_t),$$

because we have

$$\int_0^t Y(B_s,s)dB_s = Y(B_t,t) - Y(B_0,0) = Y(B_t,t) - 1.$$

The transformation $Y(B_t,t)$ of Brownian motion is sometimes called *geometric Brownian motion* or the *stochastic exponent* of the Brownian motion. It plays an important role in the celebrated Black–Scholes option pricing model [21].

5.2 Stochastic Integration with Respect to the Poisson Process

Compensated Poisson Process

The next process we look at is the Poisson process $\{N_t, t \geq 0\}$, where N_t has a Poisson distribution $P(t)$. Because $E[N_t] = t \neq 0$ we work with the *compensated Poisson process* $M_t = N_t - t$. Because $E[M_t] = 0$ the results of our stochastic integrals are martingales. For this compensated Poisson process we have a similar result. The orthogonal polynomials with respect to the Poisson distribution $P(t)$ are the Charlier polynomials and they have the generating function

$$\sum_{n=0}^{\infty} C_n(x;t)\frac{z^n}{n!} = e^z\left(1 - \frac{z}{t}\right)^x.$$

We have the following theorem.

Theorem 8

$$\int_{(0,t]} \tilde{C}_n(N_{s-}; s) dM_s = \frac{\tilde{C}_{n+1}(N_t; t)}{n+1}, \tag{5.3}$$

with $\tilde{C}_n(x;t)$, the monic Charlier polynomial of degree n.

Proof: We verify this theorem by a direct calculation.

Using the generating function of the monic Charlier polynomials, one can easily see that it is sufficient to prove

$$\int_0^t Y(N_{s-}, s, w) dM_s = \frac{Y(N_t, t, w) - 1}{w},$$

where

$$Y(X, t, w) = \exp(-tw)(1+w)^X = \sum_{m=0}^{\infty} \tilde{C}_m(X;t) \frac{w^m}{m!}.$$

Define τ_i the time of the ith jump in the Poisson process $\{N_t, t \geq 0\}$. For convenience we set $\tau_0 = 0$. Note that

$$N_{\tau_i -} = \lim_{s \to \tau_i, s < \tau_i} N_s = i - 1, \quad i \geq 1.$$

So we have

$$\begin{aligned}
&\int_0^t Y(N_{s-}, s, w) dM_s \\
&= \int_0^t e^{-sw}(1+w)^{N_{s-}} dN_s - \int_0^t e^{-sw}(1+w)^{N_{s-}} ds \\
&= \sum_{i=1}^{N_t} e^{-\tau_i w}(1+w)^{N_{\tau_i -}} - \sum_{i=1}^{N_t} \int_{(\tau_{i-1}, \tau_i]} e^{-sw}(1+w)^{N_{s-}} ds \\
&\quad - \int_{(\tau_{N_t}, t]} e^{-sw}(1+w)^{N_{s-}} ds \\
&= \sum_{i=1}^{N_t} e^{-\tau_i w}(1+w)^{i-1} - \sum_{i=1}^{N_t} \int_{(\tau_{i-1}, \tau_i]} e^{-sw}(1+w)^{i-1} ds \\
&\quad - \int_{(\tau_{N_t}, t]} e^{-sw}(1+w)^{N_t} ds \\
&= \sum_{i=1}^{N_t} e^{-\tau_i w}(1+w)^{i-1} + \sum_{i=1}^{N_t} (1+w)^{i-1} \left(\frac{e^{-\tau_i w} - e^{-\tau_{i-1} w}}{w} \right) \\
&\quad + (1+w)^{N_t} \left(\frac{e^{-tw} - e^{-\tau_{N_t} w}}{w} \right) \\
&= \frac{e^{-tw}(1+w)^{N_t} - 1}{w} \\
&= \frac{Y(N_t, t, w) - 1}{w}.
\end{aligned}$$

This proves the theorem. ◇

The interpretation is that the monic Charlier polynomials are the counterparts for Itô's integral of the usual powers

$$M_s^n = (N_s - s)^n = (\tilde{C}_1(N_s; s))^n, \quad n \geq 0,$$

for the compensated Poisson process $\{M_t, t \geq 0\}$. This result goes back to Ogura [87] and Engel [45].

The theorem also implies the results of Section 4.3.2; i.e., the monic Charlier polynomials $\{\tilde{C}_n(N_t; t)\}$ are martingales.

Stochastic Exponential

An alternative proof is based on the formula for the *stochastic exponential* or *Doléans-Dade exponential* of a semimartingale. For a given semimartingale $X = \{X_t, t \geq 0\}$, with $X_0 = 0$, the stochastic exponential of X, written $\mathcal{E}(X)$ is the process Z which is the unique solution of

$$Z_t = 1 + \int_{(0,t]} Z_{s-} dX_s.$$

It is given by the formula ([94], Theorem 36)

$$\mathcal{E}(X)(t) = Z_t = \exp\left(X_t - \frac{1}{2}[X,X]_t^c\right) \prod_{0<s\leq t} (1 + \Delta X_s) e^{-\Delta X_s}, \qquad (5.4)$$

with $\Delta X_t = X_t - X_{t-}$, the jump of X at time t and $[X,X]_t^c$, the path by path continuous part of the quadratic variation process (see [94], p. 58).

Using the generating function of the monic Charlier polynomials, one can easily see that it is sufficient, as above, to prove

$$Y(N_t, t, w) = 1 + w \int_{(0,t]} Y(N_{s-}, s-, w) dM_s.$$

To prove this we look at the stochastic exponential of a multiple w of the Poisson process N_t. Because wN_t is a quadratic pure jump process, we have $[N,N]_t^c = 0$ (see [94], p. 63). Furthermore we know that the process wN has exactly N_t jumps between 0 and t and these jumps are all of size w.

So we easily find, using (5.4),

$$\mathcal{E}(wN)(t) = (1 + w)^{N_t}.$$

Using Theorem 37 of [94] we find the stochastic exponential of a multiple w of the compensated Poisson process wM:

$$\mathcal{E}(wM)(t) = (1 + w)^{N_t} e^{-wt}.$$

Equivalently, this means

$$(1 + w)^{N_t} e^{-wt} = 1 + w \int_{(0,t]} (1 + w)^{N_{s-}} e^{-ws} dM_s,$$

from which Theorem 7 follows immediately.

Note that we found that

$$Y(N_t, t, 1) = 2^{N_t} e^{-t}$$

is the stochastic exponential or Doléans–Dade exponential of the compensated Poisson process $\{M_t, t \geq 0\}$.

5.3 Stochastic Summation with Respect to the Binomial Process

Binomial Process and Compensated Binomial Process

Let $X_1, X_2, X_3, \ldots$ be i.i.d. Bernoulli random variables with parameter $p = \Pr(X_i = 1)$. Denote

$$S_n = \sum_{i=1}^{n} X_i.$$

We call S_n the *binomial process* with parameter p because S_n has a binomial distribution $\text{Bin}(n, p)$. Set

$$T_n = S_n - E[S_n] = S_n - np.$$

Next we define stochastic integration (summation) with respect to this *compensated binomial process* T_n.

Stochastic Summation

We define, inspired by Stieltjes integration, the discrete analogue of the stochastic integral with respect to T_n as follows.

$$\int_0^n f(S_{i-1}, i-1) dT_i = \sum_{i=1}^{n} f(S_{i-1}, i-1) X_i - p \sum_{i=1}^{n} f(S_{i-1}, i-1).$$

Completely similar to the other two cases we now look at the monic orthogonal polynomials with respect to the binomial distribution. These polynomials are the monic Krawtchouk polynomials and are determined by the generating function

$$\sum_{m=0}^{\infty} \tilde{K}_m(x; n, p) \frac{z^m}{m!} = (1 + qz)^x (1 - pz)^{n-x},$$

where $n \in \{0, 1, 2, \ldots\}$, $0 < p < 1$, and $p + q = 1$.

Note that we define for each parameter $0 < p < 1$ and $n \geq 0$ an infinite set of polynomials, but we only have orthogonality for a finite set of these polynomials because we have only a finite set of mass points in our discrete distribution.

We prove the following result.

Theorem 9

$$\int_0^n \tilde{K}_j(S_{i-1}; i-1, p)dT_i = \frac{\tilde{K}_{j+1}(S_n; n, p)}{j+1}, \tag{5.5}$$

with $\tilde{K}_j(x; m, p)$, the monic Krawtchouk polynomial of degree j.

Proof: First denote by n_i the index where the ith jump occurs in the binomial process $\{S_n, n \geq 0\}$ and set $n_0 = 0$. This means that $S_{n_i} = i$, $S_{n_i - 1} = i - 1$, and

$$X_j = 1 \qquad \text{if and only if } j \in \{n_1, n_2, \ldots\}.$$

Using the generating function it is sufficient to prove

$$\int_0^n Y(S_{i-1}, i-1, z)dT_i = \frac{Y(S_n, n, z) - 1}{z},$$

where

$$Y(X, m, z) = (1 + qz)^X (1 - pz)^{m-X}.$$

Similar to Theorem 8 of the Poisson case,

$$\begin{aligned}
&\int_0^n Y(S_{i-1}, i-1, z)dT_i \\
&= \sum_{i=1}^n Y(S_{i-1}, i-1, z)X_i - p\sum_{i=1}^n Y(S_{i-1}, i-1, z) \\
&= \sum_{j=1}^{S_n} Y(S_{n_j - 1}, n_j - 1, z) - p\sum_{j=1}^{S_n} \sum_{i=n_{j-1}+1}^{n_j} Y(S_{i-1}, i-1, z) \\
&\qquad -p \sum_{i=n_{S_n}+1}^{n} Y(S_{i-1}, i-1, z) \\
&= \sum_{j=1}^{S_n} Y(j-1, n_j - 1, z) - p\sum_{j=1}^{S_n} \sum_{i=n_{j-1}+1}^{n_j} Y(j-1, i-1, z) \\
&\qquad -p \sum_{i=n_{S_n}+1}^{n} Y(S_n, i-1, z) \\
&= \sum_{j=1}^{S_n} \left(\frac{1+qz}{1-pz}\right)^{j-1} (1-pz)^{n_j - 1} \\
&\qquad -p\sum_{j=1}^{S_n} \sum_{i=n_{j-1}+1}^{n_j} \left(\frac{1+qz}{1-pz}\right)^{j-1} (1-pz)^{i-1} \\
&\qquad -p \sum_{i=n_{S_n}+1}^{n} \left(\frac{1+qz}{1-pz}\right)^{S_n} (1-pz)^{i-1}
\end{aligned}$$

$$
\begin{aligned}
&= \sum_{j=1}^{S_n} \left(\frac{1+qz}{1-pz}\right)^{j-1} (1-pz)^{n_j-1} \\
&\quad -p\sum_{j=1}^{S_n} \left(\frac{1+qz}{1-pz}\right)^{j-1} \sum_{i=n_{j-1}+1}^{n_j} (1-pz)^{i-1} \\
&\quad -p\left(\frac{1+qz}{1-pz}\right)^{S_n} \sum_{i=n_{S_n}+1}^{n} (1-pz)^{i-1} \\
&= \sum_{j=1}^{S_n} \left(\frac{1+qz}{1-pz}\right)^{j-1} (1-pz)^{n_j-1} \\
&\quad -p\sum_{j=1}^{S_n} \left(\frac{1+qz}{1-pz}\right)^{j-1} \frac{(1-pz)^{n_j}-(1-pz)^{n_{j-1}}}{-pz} \\
&\quad -p\left(\frac{1+qz}{1-pz}\right)^{S_n} \frac{(1-pz)^{n}-(1-pz)^{n_{S_n}}}{-pz} \\
&= \sum_{j=1}^{S_n} \left(\frac{1+qz}{1-pz}\right)^{j-1} (1-pz)^{n_j} \left(\frac{1}{1-pz}+\frac{1}{z}-\frac{1+qz}{z(1-pz)}\right) \\
&\quad +\frac{1}{z}\left(\left(\frac{1+qz}{1-pz}\right)^{S_n} (1-pz)^{n} - 1\right) \\
&= \frac{Y(S_n,n,z)-1}{z}.
\end{aligned}
$$

This proves our theorem. ◇

The interpretation is now that the monic Krawtchouk polynomials are the counterparts of the usual powers $(S_n - np)^n = (\tilde{K}_1(S_n; n, p))^n$, $n \geq 1$ for Itô's stochastic integral with respect to the compensated binomial process $\{T_n, n = 0, 1, 2, \ldots\}$. Also we found that the role of the classical exponential function now is taken by $Y(S_n, n, 1) = (1+q)^{S_n} q^{n-S_n}$ because of the relation

$$
\int_0^n \left(\frac{1+q}{q}\right)^{S_{i-1}} q^{i-1} dT_i = \left(\frac{1+q}{q}\right)^{S_n} q^n - 1.
$$

5.4 Chaotic and Predictable Representations for Lévy Processes

5.4.1 *Chaotic and Predictable Representation Property*

Assume we are given a complete probability space $(\Omega, \mathcal{F}, P)$. If X is a Lévy process, then there exists a unique modification of it which is càdlàg and which is also a Lévy process ([94], Theorem 30, p. 21). We henceforth

always assume that we are using this unique càdlàg version of any given Lévy process. If we let $\mathcal{F}_t = \mathcal{G}_t \vee \mathcal{N}$, where $\mathcal{G}_t = \sigma\{X_s; 0 \leq s \leq t\}$ is the natural filtration of X, and $\mathcal{N}$ are the P-null sets of $\mathcal{F}$, then $(\mathcal{F}_t)_{0 \leq t < \infty}$ is right continuous ([94], Theorem 31, p. 22). So we denote by $L^2(\Omega, \mathcal{F}) = L^2(\Omega, \mathcal{F}, (\mathcal{F}_t)_{0 \leq t \leq \infty}, P)$, the filtered and complete probability space of all square integrable random variables.

It is well known [112], that Brownian motion has the so-called *chaotic representation property* (CRP); i.e., if $F \in L^2(\Omega, \mathcal{F})$, then there exist deterministic functions $f_j \in L^2(\mathbb{R}^j_+)$ such that

$$F = E[F] + \sum_{j=1}^{\infty} \int_0^{\infty} \int_0^{t_1} \cdots \int_0^{t_{j-1}} f_j(t_1, \ldots, t_j) dB_{t_j} \cdots dB_{t_2} dB_{t_1}.$$

A consequence of it is the so-called *predictable representation property* (PRP), which says that if $F \in L^2(\Omega, \mathcal{F})$, there exists a predictable process φ_s such that

$$F = E[F] + \int_0^{\infty} \varphi_s dB_s.$$

Applications can be found in stochastic calculus; also PRP implies the completeness of the Black–Scholes option pricing model [21].

If we go the other way around we encounter Hermite polynomials. If we take for some $n \geq 0$, $f_n(t_1, \ldots, t_n) = 1_{((t_1, \ldots, t_n) \in [0,t]^n)}$ and for all $i \neq n$, $f_i = 0$, then we obtain a consequence of Theorem 7; i.e.,

$$\int_0^t \int_0^{t_1} \cdots \int_0^{t_{n-1}} dB_{t_j} \cdots dB_{t_2} dB_{t_1} = \frac{\tilde{H}_n(B_t; t)}{n!}.$$

Also if we take in the PRP, $F = \tilde{H}_{n+1}(B_t; t)/(n+1)$, for which $E[F] = 0$, Theorem 7 tells us that $\varphi_s = H_n(B_s; s) 1_{(s \in [0,t])}$.

The compensated Poisson process also has the CRP and PRP. If $F \in L^2(\Omega, \mathcal{F})$, then there exist deterministic functions $f_j \in L^2(\mathbb{R}^j_+)$ such that

$$F = E[F] + \sum_{j=1}^{\infty} \int_0^{\infty} \int_0^{t_1-} \cdots \int_0^{t_{j-1}-} f_j(t_1, \ldots, t_j) dM_{t_j} \cdots dM_{t_2} dM_{t_1}$$

and there exists a predictable process φ_s such that

$$F = E[F] + \int_0^{\infty} \varphi_s dM_s.$$

Obviously, taking again for some $n \geq 0$, $f_n(t_1, \ldots, t_n) = 1_{((t_1, \ldots, t_n) \in [0,t]^n)}$ and for all $i \neq n$, $f_i = 0$, we obtain

$$\int_0^t \int_0^{t_1-} \cdots \int_0^{t_{n-1}-} dM_{t_j} \cdots dM_{t_2} dM_{t_1} = \frac{\tilde{C}_n(N_t; t)}{n!}.$$

Also similar to the Brownian case, we can take in the predictable representation, $F = \tilde{C}_{n+1}(N_t;t)/(n+1)$, for which $E[F] = 0$. Theorem 8 then tells us that $\varphi_s = C_n(N_s;s)1_{(s\in[0,t])}$.

In general a stochastic process, $\{X_t, t \geq 0\}$ possesses the CRP if any square integrable random variable with respect to X can be expressed as an orthogonal sum of multiple stochastic integrals of deterministic functions with respect to X.

The CRP has been studied by Emery [44] for normal martingales, i.e., a martingale X such that $\langle X, X\rangle_t = ct$, for some constant $c > 0$. Later it was shown ([26], p. 207 and [27]), that the only normal martingales that possess the CRP and the weaker PRP, and are at the same time also Lévy processes, are Brownian motion and the compensated Poisson process.

For a general Lévy process, which has a Lévy measure with a finite Laplace transform outside the origin, we introduce the power jump processes and the related Teugels martingales. Furthermore we orthogonalize the Teugels martingales and show how their orthogonalization is closely related to classical orthogonal polynomials.

It is the intention of this section to provide a similar chaotic and predictable representation, which is now with respect to not only one but a whole sequence of related martingales: the orthogonalized Teugels martingales.

We give a chaotic representation for every square integral random variable in terms of these orthogonalized Teugels martingales. The predictable representation with respect to the same set of orthogonalized martingales of square integrable random variables and of square integrable martingales is an easy consequence of the chaotic representation. We refer to [94] for all unexplained notation. This section is based on the preprint [86].

5.4.2 Power Jump Processes and Teugels Martingales

Condition on Laplace Transform of Lévy Measure

We **suppose** that our Lévy measure satisfies for every $\varepsilon > 0$,

$$\int_{(-\varepsilon,\varepsilon)^c} \exp(\lambda|x|)\nu(dx) < \infty, \qquad \text{for some } \lambda > 0.$$

This implies that our Lévy measure satisfies

$$\int_{-\infty}^{+\infty} |x|^i \nu(dx) < \infty, \qquad i = 2, 3, \ldots, \tag{5.6}$$

and that the characteristic function $E[\exp(iuX_t)]$ is analytic in the neighborhood of 0, as such X_t has moments of all order and the polynomials will be dense in $L^2(R, d\varphi_t(x))$, where $\varphi_t(x) = \Pr(X_t \leq x)$. The analyticity of

$E[\exp(iuX_t)]$ can be seen from the fact:

$$\log E[\exp(zX_t)] = t\left(az + \frac{\sigma^2}{2}z^2 + \sum_{n=2}^{\infty}\frac{z^n}{n!}\int_{(|x|<1)} x^n \nu(dx) + \sum_{n=1}^{\infty}\frac{z^n}{n!}\int_{(|x|\geq 1)} x^n \nu(dx)\right).$$

Power Jump Processes

The following transformations of our Lévy process play an important role in the analysis. We denote

$$X_t^{(i)} = \sum_{0<s\leq t} (\Delta X_s)^i, \quad i = 2, 3, \ldots.$$

For convenience we put $X_t^{(1)} = X_t$. But note that **not necessarily** $X_t = \sum_{0<s\leq t} \Delta X_s$ holds; it is only true in the bounded variation case with $d = 0$ in (5.12). If $\sigma^2 = 0$, clearly $[X, X]_t = X_t^{(2)}$. The processes $X^{(i)} = \{X_t^{(i)}, t \geq 0\}$, $i = 1, 2, \ldots$, are again Lévy processes and we call them the *power jump processes*. They jump at the same time points as the original Lévy process.

Teugels Martingales

Because $E[X_t] = E[X_t^{(1)}] = tm_1 < \infty$ and we have ([94], p. 29) that

$$E\left[X_t^{(i)}\right] = E\left[\sum_{0<s\leq t} (\Delta X_s)^i\right] = t\int_{-\infty}^{\infty} x^i \nu(dx) = m_i t < \infty, \quad i = 2, 3, \ldots,$$

we denote by

$$Y_t^{(i)} = X_t^{(i)} - E\left[X_t^{(i)}\right] = X_t^{(i)} - m_i t, \qquad i = 1, 2, 3, \ldots,$$

the compensated power jump process of order i. $Y^{(i)}$ is a normal martingale, which the author calls after his scientific mentor the *Teugels martingale of order i*.

Remark In the case of a *Poisson process*, all power jump processes will be the same, and equal to the original Poisson process. In the case of a *Brownian motion*, all power jump processes of order greater than two will be equal to zero.

5.4.3 Strong Orthogonalization of the Teugels Martingales

An important question is the orthogonalization of the set $\{Y^{(i)}, i = 1, 2, \ldots\}$ of martingales. The space $\mathcal{M}^2$ is the space of square integrable martingales M such that $\sup_t E(M_t^2) < \infty$, and $M_0 = 0$ a.s. Notice that if $M \in \mathcal{M}^2$, then $\lim_{t\to\infty} E(M_t^2) = E(M_\infty^2) < \infty$, and $M_t = E[M_\infty|\mathcal{F}_t]$. Thus each $M \in \mathcal{M}^2$ can be identified with its terminal value M_∞.

Strong and Weak Orthogonality of Martingales

As in Protter ([94], p. 148), we say that two martingales $M, N \in \mathcal{M}^2$ are *strongly orthogonal,* we denote this by $M \times N$, if and only if their product MN is a uniformly integrable martingale. As noted in [94], p. 148, one can prove that $M \times N$ if and only if $[M, N]$ is a uniformly integrable martingale. We say that two random variables $X, Y \in L^2(\Omega, \mathcal{F})$ are *weakly orthogonal,* $X \perp Y$, if $E[XY] = 0$. Clearly, strong orthogonality implies weak orthogonality. Two subsets $\mathcal{A}, \mathcal{B} \subset L^2(\Omega, \mathcal{F})$, are weakly orthogonal and we denote this by $\mathcal{A} \perp \mathcal{B}$, if all elements of $\mathcal{A}$ are weakly orthogonal to each element of $\mathcal{B}$.

So we are looking for a set of pairwise strongly orthogonal martingales, $\{Z^{(i)}, i = 1, 2, \ldots\}$, such that each $Z^{(i)}$ is a linear combination of the $Y^{(i)}$, $j = 1, 2, \ldots, i$ such that $Z^{(i)} \times Z^{(j)}$, for $i \neq j$. It turns out that we can take for the $Z^{(i)}$ a linear combination of the $Y^{(j)}, j = 1, 2, \ldots, i$. If we set

$$Z^{(i)} = Y^{(i)} + a_{i,i-1}Y^{(i-1)} + a_{i,i-2}Y^{(i-2)} + \ldots + a_{i,1}Y^{(1)}, \qquad i = 1, 2, \ldots,$$

we have that

$$\left[Z^{(i)}, Y^{(j)}\right] = X^{(i+j)} + a_{i,i-1}X^{(i+j-1)} + \ldots + a_{i,1}X^{(1+j)} + \sigma^2 t 1_{(j=1)}$$

and thus $\left[Z^{(i)}, Y^{(j)}\right]$ is a martingale if and only if we have $E\left[Z^{(i)}, Y^{(j)}\right]_1 = 0$.

Orthogonalization by Isometry

Consider two spaces. The first space S_1 is the space of all real polynomials on the positive real line endowed with a scalar product $\langle ., . \rangle_1$, given by

$$\langle P(x), Q(x) \rangle_1 = \int_{-\infty}^{+\infty} P(x)Q(x)x^2\nu(dx) + \sigma^2 P(0)Q(0).$$

Note that

$$\langle x^{i-1}, x^{j-1} \rangle_1 = \int_{-\infty}^{+\infty} x^{i+j}\nu(dx) = m_{i+j} + \sigma^2 1_{(i=j=1)} < \infty, \ i, j = 1, 2, \ldots.$$

The other space S_2 is the space of all linear transformations of the Teugels martingales of the Lévy process; i.e.,

$$S_2 = \left\{a_1 Y^{(1)} + a_2 Y^{(2)} + \ldots + a_n Y^{(n)}; n \in \{1, 2, \ldots\}, a_i \in R, i = 1, \ldots, n\right\}.$$

We endow this space with the scalar product $\langle ., . \rangle_2$, given by

$$\begin{aligned}\langle Y^{(i)}, Y^{(j)} \rangle_2 &= E\left[\left[Y^{(i)}, Y^{(j)}\right]_1\right] \\ &= E\left[X_1^{(i+j)}\right] + \sigma^2 1_{(i=j=1)} = m_{i+j} + \sigma^2 1_{(i=j=1)}, \quad i, j = 1, 2, \ldots.\end{aligned}$$

So one clearly sees that $x^{i-1} \longleftrightarrow Y^{(i)}$ is an isometry between S_1 and S_2. An orthogonalization of $\{1, x, x^2, \ldots\}$ in S_1 gives an orthogonalization of $\{Y^{(1)}, Y^{(2)}, Y^{(3)}, \ldots\}$. In the examples some well-known orthogonal polynomials, like the Laguerre, Meixner, and Meixner–Pollaczek polynomials, turn up in this context.

In order to identify the exact coefficients in the orthogonalization procedure we proceed as follows. Let $\{P_n(x), n \geq 0\}$ be a system of orthogonal polynomials with respect to the inner product without the jump of σ^2 in zero; i.e.,

$$\langle f, g \rangle_3 = \int_{-\infty}^{+\infty} f(x) g(x) x^2 \nu(dx).$$

If we write

$$K_n(x) = \sum_{i=0}^{n} \frac{P_i(x) P_i(0)}{\langle P_i, P_i \rangle_3}, \qquad n = 0, 1, \ldots$$

for the so-called *kernel polynomials* of $\{P_n(x), n \geq 0\}$ [25], then following [14], a system of orthogonal polynomials, denoted by $\{P_n^{\sigma^2}(x), n \geq 0\}$ with respect to $\langle ., . \rangle_1$, is given by

$$P_n^{\sigma^2}(x) = (1 + \sigma^2 K_{n-1}(0)) P_n(x) - \sigma^2 P_n(0) K_{n-1}(x), \qquad n = 0, 1, 2, \ldots.$$

A consequence is that the $P_n^{\sigma^2}(x)$ can be obtained by using connection coefficients. Indeed in [14] it is shown that one has:

$$P_n^{\sigma^2}(x) = (1 + \sigma^2 q_{n,n}) P_n(x) - \sigma^2 \sum_{k=0}^{n-1} q_{n,k} P_k(x), \qquad n = 0, 1, 2, \ldots,$$

where

$$q_{n,k} = \frac{P_n(0) P_k(0)}{\langle P_k, P_k \rangle_3} \quad \text{and} \quad q_{n,n} = K_{n-1}(0) = \sum_{i=0}^{n-1} \frac{(P_i(0))^2}{\langle P_i, P_i \rangle_3}, \quad 0 \leq k \leq n.$$

Furthermore it is quite often relevant that the system of kernel polynomials $\{K_n(x), n \geq 0\}$ is orthogonal with respect to the measure $x^3 \nu(dx)$ [25].

5.4.4 Representation Properties

Representation of a Lévy Process Power

We express $(X_{t+t_0} - X_{t_0})^k, t_0, t \geq 0, k = 1, 2, 3, \ldots$, as a sum of stochastic integrals with respect to the special processes $Y^{(j)}, j = 1, \ldots, k$. Indeed, for $k = 1$, we have $(X_{t+t_0} - X_{t_0}) = \int_{t_0}^{t_0+t} dX_s = \int_{t_0}^{t_0+t} dY_s^{(1)} + m_1 \int_{t_0}^{t_0+t} ds = \int_{t_0}^{t_0+t} dY_s^{(1)} + m_1 t$.

Using Ito's Formula ([94], Theorem 33, p. 74), we can write for $k \geq 2$,

$$
\begin{aligned}
&(X_{t+t_0} - X_{t_0})^k \\
&= \int_0^t k(X_{(s+t_0)-} - X_{t_0})^{k-1} d(X_{s+t_0} - X_{t_0}) \\
&\quad + \frac{\sigma^2}{2} \int_0^t k(k-1)(X_{(s+t_0)-} - X_{t_0})^{k-2} ds \\
&\quad + \sum_{0<s\leq t} [(X_{s+t_0} - X_{t_0})^k - (X_{(s+t_0)-} - X_{t_0})^k \\
&\quad - k(X_{(s+t_0)-} - X_{t_0})^{k-1} \Delta X_{s+t_0}] \\
&= \int_{t_0}^{t_0+t} k(X_{u-} - X_{t_0})^{k-1} dX_u^{(1)} \\
&\quad + \frac{\sigma^2}{2} k(k-1) \left((X_{t+t_0} - X_{t_0})^{k-2}(t+t_0) - \int_0^t s d(X_{s+t_0} - X_{t_0})^{k-2} \right) \\
&\quad + \sum_{0<s\leq t} [(X_{(s+t_0)-} + \Delta X_{s+t_0} - X_{t_0})^k - (X_{(s+t_0)-} - X_{t_0})^k \\
&\quad - k(X_{(s+t_0)-} - X_{t_0})^{k-1} \Delta X_{s+t_0}] \\
&= \int_{t_0}^{t_0+t} k(X_{u-} - X_{t_0})^{k-1} dX_u^{(1)} \\
&\quad + \frac{\sigma^2}{2} k(k-1) \left((X_{t+t_0} - X_{t_0})^{k-2}(t+t_0) - \int_0^t s d(X_{s+t_0} - X_{t_0})^{k-2} \right) \\
&\quad + \sum_{0<s\leq t} \sum_{j=2}^{k} \binom{k}{j} (X_{(s+t_0)-} - X_{t_0})^{k-j} (\Delta X_{s+t_0})^j \\
&= \int_{t_0}^{t_0+t} k(X_{u-} - X_{t_0})^{k-1} dX_u^{(1)} \\
&\quad + \frac{\sigma^2}{2} k(k-1) \left((X_{t+t_0} - X_{t_0})^{k-2}(t+t_0) - \int_0^t s d(X_{s+t_0} - X_{t_0})^{k-2} \right) \\
&\quad + \sum_{t_0<u\leq t+t_0} \sum_{j=2}^{k} \binom{k}{j} (X_{u-} - X_{t_0})^{k-j} (\Delta X_u)^j \\
&= \int_{t_0}^{t_0+t} k(X_{u-} - X_{t_0})^{k-1} dX_u^{(1)} \\
&\quad + \frac{\sigma^2}{2} k(k-1) \left((X_{t+t_0} - X_{t_0})^{k-2}(t+t_0) - \int_0^t s d(X_{s+t_0} - X_{t_0})^{k-2} \right) \\
&\quad + \sum_{j=2}^{k} \binom{k}{j} \int_{t_0}^{t+t_0} (X_{u-} - X_{t_0})^{k-j} dX_u^{(j)}. \qquad (5.7)
\end{aligned}
$$

Next we prove that the power of an increment of a Lévy process, $(X_{t+t_0} - X_{t_0})^k$, has a representation of the form

$$(X_{t+t_0} - X_{t_0})^k = f^{(k)}(t,t_0) + \sum_{j=1}^{k} \sum_{1 \le i_1,\dots,i_j \le k} \int_{t_0}^{t+t_0} \int_{t_0}^{t_1-} \dots \int_{t_0}^{t_{j-1}-} f^{(k)}_{(i_1,\dots,i_j)} dY^{(i_j)}_{t_j} \dots dY^{(i_2)}_{t_2} dY^{(i_1)}_{t_1}, \tag{5.8}$$

where the $f^{(k)}_{(i_1,\dots,i_j)} = f^{(k)}_{(i_1,\dots,i_j)}(t,t_0,t_1,\dots,t_j)$ are real deterministic functions. Moreover from what follows it can be seen that they are just real multivariate polynomials of total degree less than k and that we have $f^{(k)}_{(i_1,\dots,i_j)} = 0$, whenever $i_1 + \dots + i_j > k$.

The representation follows from (5.7), where we bring in the right compensators; i.e., we can write

$$\begin{aligned}
&\sum_{j=1}^{k} \binom{k}{j} \int_{t_0}^{t+t_0} (X_{s-} - X_{t_0})^{k-j} dX_s^{(j)} \\
&= \sum_{j=1}^{k} \binom{k}{j} \int_{t_0}^{t+t_0} (X_{s-} - X_{t_0})^{k-j} dY_s^{(j)} \\
&\quad + \sum_{j=1}^{k} \binom{k}{j} m_j \int_{t_0}^{t+t_0} (X_{s-} - X_{t_0})^{k-j} ds \\
&= \sum_{j=1}^{k} \binom{k}{j} \int_{t_0}^{t+t_0} (X_{s-} - X_{t_0})^{k-j} dY_s^{(j)} \\
&\quad + \sum_{j=1}^{k-1} \binom{k}{j} m_j (t+t_0)(X_{t+t_0} - X_{t_0})^{k-j} \\
&\quad - \sum_{j=1}^{k-1} \binom{k}{j} m_j \int_{t_0}^{t+t_0} s d(X_s - X_{t_0})^{k-j} + m_k t.
\end{aligned} \tag{5.9}$$

Combining (5.7) and (5.9) gives

$$\begin{aligned}
&(X_{t+t_0} - X_{t_0})^k \\
&= \frac{\sigma^2}{2} k(k-1) \left((X_{t+t_0} - X_{t_0})^{k-2}(t+t_0) - \int_0^t s d(X_{s+t_0} - X_{t_0})^{k-2} \right) \\
&\quad + \sum_{j=1}^{k} \binom{k}{j} \int_{t_0}^{t+t_0} (X_{s-} - X_{t_0})^{k-j} dY_s^{(j)} \\
&\quad + \sum_{j=1}^{k-1} \binom{k}{j} m_j (t+t_0)(X_{t+t_0} - X_{t_0})^{k-j}
\end{aligned}$$

$$-\sum_{j=1}^{k-1}\binom{k}{j}m_j\int_{t_0}^{t+t_0} sd(X_s - X_{t_0})^{k-j} + m_k t. \tag{5.10}$$

The last equation is in terms of powers of increments of X that are strictly lower than k. So by induction the representation (5.8) can be proven.

Lemma 1 *We have that*

$$E[(X_{t+t_0} - X_{t_0})^k] = f^{(k)}(t, t_0) = f^{(k)}(t), \qquad t, t_0 \geq 0,$$

which is independent of t_0.

Proof: This follows from the fact that stochastic integrals with respect to martingales are martingales and thus have mean zero, and that a Lévy process has stationary increments, thus the mean of $(X_{t+t_0} - X_{t_0})^k$ is not dependent on t_0.◇

As an illustration we give the $f^{(k)}_{(i_1,\ldots,i_j)}$ for $k = 0, 1, 2$ and $t_0 = 0$ and $\sigma^2 = 0$. We start with the trivial case $k = 0$. Here we have $X_t^0 = 1$, so that $f^{(0)}(t) = f^{(0)}(t, 0) = 1$. Also the case $k = 1$ is easy. Indeed, because $X_t^1 = Y_t^{(1)} + m_1 t = \int_0^t dY_{t_1}^{(1)} + m_1 t$, we clearly see that $f^{(1)}(t, 0) = f^{(1)}(t) = m_1 t$ and $f^{(1)}_{(1)}(t, 0, t_1) = 1$. The case $k = 2$ is a little more complex. We start from (5.10):

$$\begin{aligned}
& X_t^2 \\
&= 2\int_0^t X_{t_1-}dY_{t_1}^{(1)} + \int_0^t dY_{t_1}^{(2)} + 2m_1\left(tX_t - \int_0^t t_1 dX_{t_1}\right) + m_2 t \\
&= 2\int_0^t \left(\int_0^{t_1-} dY_{t_2}^{(1)} + m_1 t_1\right) dY_{t_1}^{(1)} + \int_0^t dY_{t_1}^{(2)} \\
&\quad +2m_1\left(tX_t - \int_0^t t_1 dX_{t_1}\right) + m_2 t \\
&= 2\int_0^t\int_0^{t_1-} dY_{t_2}^{(1)}dY_{t_1}^{(1)} + 2m_1\int_0^t t_1 dY_{t_1}^{(1)} + \int_0^t dY_{t_1}^{(2)} \\
&\quad +2m_1 t\int_0^t dY_{t_1}^{(1)} + 2m_1^2t^2 - 2m_1\int_0^t t_1 dY_{t_1}^{(1)} - 2m_1^2\int_0^t t_1 dt_1 + m_2 t \\
&= 2\int_0^t\int_0^{t_1-} dY_{t_2}^{(1)}dY_{t_1}^{(1)} + \int_0^t dY_{t_1}^{(2)} + 2m_1 t\int_0^t dY_{t_1}^{(1)} + m_1^2t^2 + m_2 t.
\end{aligned}$$

So that

$$\begin{aligned}
f^{(2)}(t,0) &= m_1^2t^2 + m_2 t, \\
f^{(2)}_{(1)}(t,0,t_1) &= 2m_1 t, \qquad f^{(2)}_{(2)}(t,0,t_1) = 1, \\
f^{(2)}_{(1,1)}(t,0,t_1,t_2) &= 2, \qquad f^{(2)}_{(1,2)}(t,0,t_1,t_2) = f^{(2)}_{(2,2)}(t,0,t_1,t_2) = 0.
\end{aligned}$$

Because we can switch by a linear transformation from the $Y^{(i)}$ to the $Z^{(i)}$, it is clear that we also have a representation of the form

$$(X_{t+t_0} - X_{t_0})^k = f^{(k)}(t) + \sum_{j=1}^{k} \sum_{1 \le i_1,\ldots,i_j \le k} \int_{t_0}^{t+t_0} \int_{t_0}^{t_1-} \cdots \int_{t_0}^{t_{j-1}-} h^{(k)}_{(i_1,\ldots,i_j)} dZ^{(i_j)}_{t_j} \ldots dZ^{(i_2)}_{t_2} dZ^{(i_1)}_{t_1}, \tag{5.11}$$

where the $h^{(k)}_{(i_1,\ldots,i_j)} = h^{(k)}_{(i_1,\ldots,i_j)}(t, t_0, t_1, \ldots, t_j)$ are real deterministic functions in $L^2(\mathbb{R}^j)$.

Representation of a Square Integrable Random Variable

We denote by

$$\mathcal{H}^{(i_1,\ldots,i_j)} = \{F \in L^2(\Omega, \mathcal{F});$$
$$F = \textstyle\int_0^\infty \int_0^{t_1-} \cdots \int_0^{t_{j-1}-} f(t_1,\ldots,t_j) dZ^{(i_j)}_{t_j} \ldots dZ^{(i_2)}_{t_2} dZ^{(i_1)}_{t_1}, f \in L^2(\mathbb{R}^j)\}.$$

We say that two multi-indexes $(i_1,\ldots,i_k)$ and $(j_1,\ldots,j_l)$ are different if $k \neq l$ or when $k = l$, if there exists a subindex $1 \le n \le k = l$, such that $i_n \neq j_n$, and denote this by

$$(i_1,\ldots,i_k) \neq (j_1,\ldots,j_l).$$

Proposition 2 *If* $(i_1,\ldots,i_k) \neq (j_1,\ldots,j_l)$*, then* $\mathcal{H}^{(i_1,\ldots,i_k)} \perp \mathcal{H}^{(j_1,\ldots,j_l)}$.

Proof: Suppose we have two random variables $K \in \mathcal{H}^{(i_1,\ldots,i_k)}$ and $L \in \mathcal{H}^{(j_1,\ldots,j_l)}$. We need to prove that if $(i_1,\ldots,i_k) \neq (j_1,\ldots,j_l)$, then $K \perp L$.

For the case $k = l$, we use induction on k. So take first $k = l = 1$. We may assume the following representations for K and L,

$$K = \int_0^\infty f(t_1) dZ^{(i_1)}_{t_1}, \qquad L = \int_0^\infty g(t_1) dZ^{(j_1)}_{t_1},$$

where we must have $i_1 \neq j_1$. By construction $Z^{(i_1)}$ and $Z^{(j_1)}$ are strongly orthogonal martingales. Using the fact that stochastic integrals w.r.t. strongly orthogonal martingales are again strongly orthogonal ([94], Lemma 2 and Theorem 35, p. 149) and thus also weakly orthogonal, it immediately follows that $K \perp L$.

Suppose the theorem holds for all $1 \le k = l \le n-1$. We prove the theorem for $k = l = n$. Assume the following representations.

$$\begin{aligned} K &= \int_0^\infty \int_0^{t_1-} \cdots \int_0^{t_{n-1}-} f(t_1,\ldots,t_n) dZ^{(i_n)}_{t_n} \ldots dZ^{(i_2)}_{t_2} dZ^{(i_1)}_{t_1} \\ &= \int_0^\infty \alpha_{t_1} dZ^{(i_1)}_{t_1}, \\ L &= \int_0^\infty \int_0^{t_1-} \cdots \int_0^{t_{n-1}-} g(t_1,\ldots,t_n) dZ^{(j_n)}_{t_n} \ldots dZ^{(j_2)}_{t_2} dZ^{(j_1)}_{t_1} \\ &= \int_0^\infty \beta_{t_1} dZ^{(j_1)}_{t_1}. \end{aligned}$$

There are two possibilities: $i = i_1 = j_1$ and $i_1 \neq j_1$. In the former case we must have that $(i_2, \ldots, i_n) \neq (j_2, \ldots, j_n)$, and thus by induction $\alpha_{t_1} \perp \beta_{t_1}$. Therefore,

$$\begin{aligned} E[KL] &= E\left[\left(\int_0^\infty \alpha_s dZ_s^{(i)}\right)\left(\int_0^\infty \beta_s dZ_s^{(i)}\right)\right] \\ &= E\left[\int_0^\infty \alpha_s \beta_s d\langle Z^{(i)}, Z^{(i)}\rangle_s\right] = \int_0^\infty E(\alpha_s \beta_s) d\langle Z^{(i)}, Z^{(i)}\rangle_s = 0. \end{aligned}$$

In the latter case we again use the fact that stochastic integrals w.r.t. strongly orthogonal martingales are again strongly orthogonal ([94], Lemma 2 and Theorem 35, p. 149) and thus also weakly orthogonal. So it immediately follows that $K \perp L$.

For the case $k \neq l$, a similar argument can be used together with the fact that all elements of every $\mathcal{H}^{(i_1,\ldots,i_n)}$, $n \geq 1$ have mean zero and thus are orthogonal w.r.t. the constants.◇

Proposition 3 *Let*

$$\begin{aligned} \mathcal{P} = \{&X_{t_1}^{k_1}(X_{t_2} - X_{t_1})^{k_2} \ldots (X_{t_n} - X_{t_{n-1}})^{k_n} : \\ &n \geq 1, 0 \leq t_1 < t_2 < \ldots < t_n, k_1, \ldots, k_n \in \{0, 1, \ldots\}\}; \end{aligned}$$

then we have that $\mathcal{P}$ is a total family in $L^2(\Omega, \mathcal{F})$; i.e., the linear subspace spanned by $\mathcal{P}$ is dense in $L^2(\Omega, \mathcal{F})$.

Proof: Let $Z \in L^2(\Omega, \mathcal{F})$ and $Z \perp \mathcal{P}$. For any given $\varepsilon > 0$, there exists a finite set of time points $\{s_1, \ldots, s_m\}$ and a random variable $Z_\varepsilon \in L^2(\Omega, \sigma(X_{s_1}, X_{s_2}, \ldots, X_{s_m}))$ such that

$$E\left[(Z - Z_\varepsilon)^2\right] < \varepsilon.$$

So there exists a Borel function f such that

$$Z_\varepsilon = f_\varepsilon(X_{s_1}, X_{s_2} - X_{s_1}, \ldots, X_{s_m} - X_{s_{m-1}}).$$

Because the characteristic function $E[\exp(iuX_t)]$ is analytic in the neighborhood of 0, the polynomials are dense in $L^2(R, d\varphi_t(x))$, and we can approximate Z_ε by polynomials. Furthermore because $Z \perp \mathcal{P}$, we have for all $k_1, k_2, \ldots, k_m \geq 0$,

$$0 = E\left[Z X_{s_1}^{k_1}(X_{s_2} - X_{s_1})^{k_2} \ldots (X_{s_m} - X_{s_{m-1}})^{k_m}\right];$$

thus $E[ZZ_\varepsilon] = 0$. Then

$$E[Z^2] = E[Z(Z - Z_\varepsilon)] \leq \sqrt{E[Z^2]E[(Z - Z_\varepsilon)^2]} \leq \sqrt{\varepsilon E[Z^2]},$$

and thus because always $0 \leq E[Z^2]$, we must have $E[Z^2] \leq \varepsilon$. Letting $\varepsilon \to 0$ yields $\varepsilon = 0$ a.s. Thus $\mathcal{P}$ is a total family in $L^2(\Omega, \mathcal{F})$.◇

Chaotic Representation Property

We are now in a position to prove our main theorem.

Theorem 10 (Chaotic Representation Property (CRP)) *Every random variable F in $L^2(\Omega, \mathcal{F})$ has a representation of the form*

$$F = E[F] + \sum_{j=1}^{\infty} \sum_{1 \le i_1, \dots, i_j \le \infty} \int_0^{\infty} \int_0^{t_1-} \cdots \int_0^{t_{j-1}-} f_{(i_1,\dots,i_j)} dZ_{t_j}^{(i_j)} \dots dZ_{t_2}^{(i_2)} dZ_{t_1}^{(i_1)},$$

where the $f_{(i_1,\dots,i_j)} = f_{(i_1,\dots,i_j)}(t_1, \dots, t_j)$ are real deterministic functions in $L^2(\mathbb{R}^j)$.

Proof: Because $\mathcal{P}$ is a total family in $L^2(\Omega, \mathcal{F})$, it is sufficient to prove that every element of $\mathcal{P}$ has a representation of the desired form. This follows from the fact that $\mathcal{P}$ is built up from terms of the form $X_{t_1}^{k_1}(X_{t_2} - X_{t_1})^{k_2} \dots (X_{t_n} - X_{t_{n-1}})^{k_n}$, wherein each term has a representation of the form based on (5.11). And we can nicely combine two terms in the desired representation. Indeed we have for all $k, l \ge 1$, and $0 \le t < s \le u < v$, that the product of $(X_s - X_t)^k (X_v - X_u)^l$ is a sum of products of the form AB where

$$A = \int_t^s \int_t^{t_1-} \cdots \int_t^{t_{n-1}-} h_{(i_1,\dots,i_n)}^{(k)}(s, t, t_1, \dots, t_n) dZ_{t_n}^{(i_n)} \dots dZ_{t_2}^{(i_2)} dZ_{t_1}^{(i_1)},$$

$$B = \int_u^v \int_u^{u_1-} \cdots \int_u^{u_{m-1}-} h_{(j_1,\dots,j_m)}^{(l)}(v, u, u_1, \dots, u_m) dZ_{u_m}^{(j_m)} \dots dZ_{u_2}^{(j_2)} dZ_{u_1}^{(j_1)}.$$

Because

$$\begin{aligned}
AB &= \int_t^s \int_t^{t_1-} \cdots \int_t^{t_{n-1}-} h_{(i_1,\dots,i_n)}^{(k)} dZ_{t_n}^{(i_n)} \dots dZ_{t_2}^{(i_2)} dZ_{t_1}^{(i_1)} \times \\
&\quad \int_u^v \int_u^{u_1-} \cdots \int_u^{u_{m-1}-} h_{(j_1,\dots,j_m)}^{(l)} dZ_{u_m}^{(j_m)} \dots dZ_{u_2}^{(j_2)} dZ_{u_1}^{(j_1)} \\
&= \int_u^v \int_u^{u_1-} \cdots \int_u^{u_{m-1}-} \int_t^s \int_t^{t_1-} \cdots \int_t^{t_{n-1}-} \\
&\quad h_{(j_1,\dots,j_m)}^{(l)} h_{(i_1,\dots,i_n)}^{(k)} dZ_{t_n}^{(i_n)} \dots dZ_{t_2}^{(i_2)} dZ_{t_1}^{(i_1)} dZ_{u_m}^{(j_m)} \dots dZ_{u_2}^{(j_2)} dZ_{u_1}^{(j_1)} \\
&= \int_0^{\infty} \int_0^{u_1-} \cdots \int_0^{u_{m-1}-} \int_0^{u_m-} \int_0^{t_1-} \cdots \int_0^{t_{n-1}-} h_{(j_1,\dots,j_m)}^{(l)} h_{(i_1,\dots,i_n)}^{(k)} \times \\
&\quad 1_{(u<u_m<\dots<u_1\le v)} 1_{(t<t_n<\dots<t_1\le s)} dZ_{t_n}^{(i_n)} \dots dZ_{t_2}^{(i_2)} dZ_{t_1}^{(i_1)} dZ_{u_m}^{(j_m)} \dots dZ_{u_2}^{(j_2)} dZ_{u_1}^{(j_1)},
\end{aligned}$$

where we wrote for simplicity $h_{(i_1,\dots,i_n)}^{(k)} = h_{(i_1,\dots,i_n)}^{(k)}(s, t, t_1, \dots, t_n)$ and $h_{(j_1,\dots,j_m)}^{(l)} = h_{(j_1,\dots,j_m)}^{(l)}(v, u, u_1, \dots, u_m)$.

Clearly, the representation follows.◇

Predictable Representation Property

Theorem 11 (Predictable Representation Property (PRP)) *Every random variable F in $L^2(\Omega, \mathcal{F})$ has a representation of the form*

$$F = E[F] + \sum_{i=1}^{\infty} \int_0^{\infty} \phi_s^{(i)} dZ_s^{(i)},$$

where $\phi_s^{(i)}$ is predictable.

Proof: From the theorem above, we know that F has a representation of the form

$$\begin{aligned}
& F - E[F] \\
&= \sum_{j=1}^{\infty} \sum_{1 \le i_1, \ldots, i_j < \infty} \int_0^{\infty} \int_0^{t_1-} \cdots \int_0^{t_{j-1}-} f_{(i_1,\ldots,i_j)} dZ_{t_j}^{(i_j)} \ldots dZ_{t_2}^{(i_2)} dZ_{t_1}^{(i_1)} \\
&= \sum_{i_1=1}^{\infty} \int_0^{\infty} f_{(i_1)}(t_1) dZ_{t_1}^{(i_1)} + \\
& \sum_{i_1=1}^{\infty} \int_0^{\infty} \left[\sum_{j=2}^{\infty} \sum_{1 \le i_2, \ldots, i_j < \infty} \int_0^{t_1-} \cdots \int_0^{t_{j-1}-} f_{(i_1,\ldots,i_j)} dZ_{t_j}^{(i_j)} \ldots dZ_{t_2}^{(i_2)} \right] dZ_{t_1}^{(i_1)} \\
&= \sum_{i_1=1}^{\infty} \int_0^{\infty} [f_{(i_1)}(t_1) + \\
& \qquad \sum_{j=2}^{\infty} \sum_{1 \le i_2, \ldots, i_j < \infty} \int_0^{t_1-} \cdots \int_0^{t_{j-1}-} f_{(i_1,\ldots,i_j)} dZ_{t_j}^{(i_j)} \ldots dZ_{t_2}^{(i_2)}] dZ_{t_1}^{(i_1)} \\
&= \sum_{i=1}^{\infty} \int_0^{\infty} \phi_{t_1}^{(i)} dZ_{t_1}^{(i)},
\end{aligned}$$

where as always $f_{(i_1,\ldots,i_j)} = f_{(i_1,\ldots,i_j)}(t_1, \ldots, t_j)$ and which is exactly of the form we want.◇

Predictable Martingale Representation

Because we can identify every martingale $M \in \mathcal{M}^2$ with its terminal value $M_\infty \in L^2(\Omega, \mathcal{F})$ and because $M_t = E[M_\infty | \mathcal{F}_t]$, we have the predictable representation

$$M_t = \sum_{i=1}^{\infty} \int_0^t \phi_s^{(i)} dZ_s^{(i)},$$

which is a sum of strongly orthogonal martingales.

Space Decomposition

Another consequence of the chaotic representation property is the following theorem.

Theorem 12 *We have the following space decomposition*

$$L^2(\Omega, \mathcal{F}) = R \bigoplus \left(\bigoplus_{j=1}^{\infty} \bigoplus_{(i_1,\dots,i_j)\in N^j} \mathcal{H}^{(i_1,\dots,i_j)} \right).$$

Remark The Lévy–Khintchine formula has a simpler expression when the sample paths of the related Lévy process have bounded variation on every compact time interval a.s. It is well known ([16], p. 15) that a Lévy process has bounded variation if and only if $\sigma^2 = 0$, and $\int_{-\infty}^{+\infty}(1 \wedge |x|)\nu(dx) < \infty$. In that case the characteristic exponent can be re-expressed as

$$\psi(\theta) = id\theta + \int_{-\infty}^{+\infty} (\exp(i\theta x) - 1)\nu(dx).$$

Furthermore we can write

$$X_t = dt + \sum_{0<s\leq t} \Delta X_s, \qquad t \geq 0, \tag{5.12}$$

and the calculations simplify somewhat because $\sigma^2 = 0$ and for $k \geq 1$,

$$\int_0^t (X_{(s+t_0)-} - X_{t_0})^{k-1} d(X_{s+t_0} - X_{t_0}) = \sum_{0<s\leq t} (X_{(s+t_0)-} - X_{t_0})^{k-1} \Delta X_{s+t_0}.$$

5.5 Examples

5.5.1 The Gamma Process

As before, the Gamma process is the Lévy process (of bounded variation), $G = \{G_t, t \geq 0\}$, with Lévy measure given by

$$\nu(dx) = 1_{(x>0)} \frac{e^{-x}}{x} dx.$$

It was called the Gamma process because the distribution of G_t is a Gamma distribution with mean equal to t and scale parameter equal to one.

Power Jump Processes

As indicated in the above discussion, the following transformations of the Gamma process play an important role in the analysis. We denote

$$G_t^{(i)} = \sum_{0<s\leq t} (\Delta G_s)^i, \qquad i = 1, 2, 3, \dots.$$

We first calculate the Lévy measure of these power jump processes. Using the exponential formula ([16], p. 8), and the change of variable $z = x^j$, we obtain for $j = 1, 2, 3, \ldots$,

$$\begin{aligned} E\left[\exp\left(i\theta G_t^{(j)}\right)\right] &= E\left[\exp\left(i\theta \sum_{0<s\leq t} (\Delta G_s)^j\right)\right] \\ &= \exp\left(t\int_0^\infty (\exp(i\theta x^j) - 1)\frac{e^{-x}}{x}dx\right) \\ &= \exp\left(t\int_0^\infty (\exp(i\theta z) - 1)\frac{\exp(-z^{1/j})}{jz}dz\right), \end{aligned}$$

which means that the Lévy measure of $G^{(j)}$ is given by

$$\frac{\exp(-z^{1/j})}{jz}dz.$$

Teugels Martingales

Because

$$E\left[\sum_{0<s\leq t} (\Delta G_s)^i\right] = t\int_0^\infty x^i\frac{e^{-x}}{x}dx, \qquad i = 1, 2, 3, \ldots,$$

we clearly have

$$E\left[G_t^{(i)}\right] = \Gamma(i)t = (i-1)!t, \qquad i = 1, 2, 3, \ldots,$$

and thus

$$\hat{G}_t^{(i)} = G_t^{(i)} - (i-1)!t, \qquad i = 1, 2, 3, \ldots,$$

is the *Teugels martingale of order i of the Gamma process.*

Next we orthogonalize the set $\{\hat{G}^{(i)}, i = 1, 2, \ldots\}$ of martingales. So we are looking for a set of martingales,

$$\{K^{(i)} = \hat{G}^{(i)} + a_{i,i-1}\hat{G}^{(i-1)} + a_{i,i-2}\hat{G}^{(i-2)} + \ldots + a_{i,1}\hat{G}^{(1)}, i = 1, 2, \ldots\},$$

such that $K^{(i)} \times K^{(j)}$, for $i \neq j$.

The first space S_1 is, in the Gamma case, the space of all real polynomials on the positive real line endowed with a scalar product $\langle ., . \rangle_1$, given by

$$\langle P(x), Q(x)\rangle_1 = \int_0^\infty P(x)Q(x)xe^{-x}dx.$$

Note that

$$\langle x^{i-1}, x^{j-1}\rangle_1 = \int_0^\infty x^{i+j-1}e^{-x}dx = (i+j-1)!, \qquad i, j = 1, 2, 3, \ldots.$$

The other space S_2 is the space of all linear transformations of the Teugels martingales of the Gamma process; i.e.,

$$S_2 = \{a_1\hat{G}^{(1)} + a_2\hat{G}^{(2)} + \ldots + a_n\hat{G}^{(n)}; n \in \{1,2,\ldots\}, a_i \in R, i = 1,\ldots,n\}$$

endowed with the scalar product $\langle .,. \rangle_2$, given by

$$\langle \hat{G}^{(i)}, \hat{G}^{(j)} \rangle_2 = E\left[\left[\hat{G}^{(i)}, \hat{G}^{(j)}\right]_1\right] = E\left[G_1^{(i+j)}\right] = (i+j-1)!,$$

for $i,j = 1,2,3,\ldots$.

So one clearly sees that $x^{i-1} \longleftrightarrow \hat{G}^{(i)}$ is an isometry between S_1 and S_2. An orthogonalization of $\{1, x, x^2, \ldots\}$ in S_1 gives the Laguerre polynomials $L_n^{(1)}(x)$ [71], so by isometry we also find an orthogonalization of $\{\hat{G}^{(1)}, \hat{G}^{(2)}, \hat{G}^{(3)}, \ldots\}$.

5.5.2 *The Pascal Process*

The next process of bounded variation we look at is the negative binomial process, also called the Pascal process. It has a Lévy measure $\nu(du)$ given by

$$\nu(du) = d\sum_{x=1}^{\infty} \frac{q^x}{x} 1_{(u \geq x)},$$

where $0 < q < 1$. One can prove that a Lévy process, $P = \{P_t, t \geq 0\}$, with such a Lévy measure has a negative binomial distribution with characteristic function given by

$$E[\exp(i\theta P_t)] = \left(\frac{p}{1 - qe^{i\theta}}\right)^t,$$

where $p = 1 - q$.

Let us denote by $P^{(i)} = \{P_t^{(i)}, t \geq 0\}$ the corresponding power jump processes and by $Q^{(i)} = \{Q_t^{(i)}, t \geq 0\}$ the corresponding Teugels martingales.

We look for the orthogonalization of the set $\{Q^{(i)}, i = 1, 2, \ldots\}$ of martingales. The space S_1 is now the space of all real polynomials on the positive real line endowed with a scalar product $\langle .,. \rangle_1$, given by

$$\langle P(x), R(x) \rangle_1 = \sum_{x=1}^{\infty} P(x)R(x)xq^x.$$

Note that

$$\langle x^{i-1}, x^{j-1} \rangle_1 = \sum_{x=1}^{\infty} q^x x^{i+j-1}, \qquad i,j = 1,2,3,\ldots.$$

The space S_2 is now the space of all linear transformations of the Teugels martingales of the negative binomial process; i.e.,

$$S_2 = \{a_1 Q^{(1)} + a_2 Q^{(2)} + \ldots + a_n Q^{(n)}; n \in \{1, 2, \ldots, n\}, a_i \in R, i = 1, \ldots, n\},$$

and is endowed with the scalar product $\langle ., . \rangle_2$, given by

$$\langle Q^{(i)}, Q^{(j)} \rangle_2 = E\left[\left[Q^{(i)}, Q^{(j)}\right]_1\right] = E\left[P_1^{(i+j)}\right] = \sum_{x=1}^{\infty} q^x x^{i+j-1},$$

for $i, j = 1, 2, \ldots$. By construction $x^{i-1} \longleftrightarrow Q^{(i)}$ is an isometry between S_1 and S_2. An orthogonalization of $\{1, x, x^2, \ldots\}$ in S_1 gives the Meixner polynomials $M_n(x-1; 2, p)$ [71], so by isometry we also find an orthogonalization of the set $\{Q^{(1)}, Q^{(2)}, Q^{(3)}, \ldots\}$.

5.5.3 *The Meixner Process*

A Meixner process $M = \{H_t, t \geq 0\}$ is a bounded variation Lévy process based on the infinitely divisible distribution with density function given by

$$f(x; m, a) = \frac{(2\cos(a/2))^{2m}}{2\pi\Gamma(2m)} \exp(ax)|\Gamma(m + ix)|^2, \qquad x \in (-\infty, +\infty),$$

and where a is a real constant and $m > 0$. The corresponding probability distribution is the measure of orthogonality of the Meixner–Pollaczek polynomials [71]. The characteristic function of M_1 is given by

$$E[\exp(i\theta M_1)] = \left(\frac{\cos(a/2)}{\cosh((\theta - ia)/2)}\right)^{2m}.$$

In [103] its Lévy measure is calculated:

$$\nu(dx) = m\frac{\exp(ax)}{x\sinh(\pi x)}dx = m|\Gamma(1 + ix)|^2 \frac{\exp(ax)}{\pi x^2}dx.$$

Note that

$$x^2\nu(dx) = m|\Gamma(1 + ix)|^2 \frac{\exp(ax)}{\pi}dx,$$

is, up to a constant, equal to $f(x; 1, a)$. Completely similarly to the two above examples, we thus can orthogonalize the Teugels martingales for the Meixner process by isometry. The orthogonal polynomials involved are now the Meixner–Pollaczek polynomials $P_n(x; 1; a)$.

5.5.4 *Brownian–Gamma Process*

Consider a Lévy process $X = \{X_t, t \geq 0\}$, which is given by the triplet $[0, \sigma^2, 1_{(x>0)}e^{-x}x^{-1}dx]$. Thus has no deterministic part and the stochastic part consists of a Brownian motion $\{B_t, t \geq 0\}$ with parameter σ^2 and an independent pure jump part $\{G_t, t \geq 0\}$ which is called a Gamma process, because the law of G_t is a gamma distribution with mean t and scale parameter equal to one. The Gamma process is used, for example, in insurance and mathematical finance [32], [39], [40]. In the orthogonalization of the Teugels martingales of this process, we work with the above-described spaces. The first space S_1 is here the space of all real polynomials on the positive real line endowed with a scalar product $\langle .,. \rangle_1$, given by

$$\langle P(x), Q(x) \rangle_1 = \int_0^\infty P(x)Q(x)xe^{-x}dx + \sigma^2 P(0)Q(0).$$

Note that

$$\langle x^{i-1}, x^{j-1} \rangle_1 = \int_0^\infty x^{i+j-1}e^{-x}dx = (i+j-1)! + \sigma^2 1_{(i=j=1)},$$

for $i, j = 1, 2, 3, \ldots$. The other space S_2 is the space of all linear transformations of the Teugels martingales $\{Y_t^{(i)}, t \geq 0\}$ of our Lévy process X; i.e.,

$$S_2 = \{a_1 Y^{(1)} + a_2 Y^{(2)} + \ldots + a_n Y^{(n)}; n \in \{1, 2, \ldots\}, a_i \in \mathbb{R}, i = 1, \ldots, n\}$$

endowed with the scalar product $\langle .,. \rangle_2$, given by

$$\langle Y^{(i)}, Y^{(j)} \rangle_2 = (i+j-1)! + \sigma^2 1_{(i=j=1)}, \qquad i, j = 1, 2, 3, \ldots.$$

Therefore one clearly sees that $x^{i-1} \longleftrightarrow Y^{(i)}$ is an isometry between S_1 and S_2. An orthogonalization of $\{1, x, x^2, \ldots\}$ in S_1 gives the *Laguerre-type polynomials* $L_n^{1,\sigma^2}(x)$ introduced in [76], so by isometry we also find an orthogonalization of $\{Y^{(1)}, Y^{(2)}, Y^{(3)}, \ldots\}$.

Next we explicitly calculate the coefficients $\{a_{ij}, 1 \leq i \leq j\}$, such that

$$\{Z^{(j)} = a_{1j} Y^{(1)} + a_{2j} Y^{(2)} + \ldots + a_{jj} Y^{(j)}, j = 1, 2, \ldots\}$$

is a strongly pairwise orthogonal sequence of martingales. We make use of the Laguerre polynomials $\{L_n^{(\alpha)}(x), n = 0, 1, \ldots\}$ defined for every $\alpha > 1$ by

$$L_n^{(\alpha)}(x) = \frac{1}{n!} \sum_{k=0}^{n} (-n)_k (\alpha + k + 1)_{n-k} \frac{x^k}{k!}, \quad n = 0, 1, \ldots \tag{5.13}$$

and their kernel polynomials $\{K_n(x), n \geq 0\}$. The Laguerre polynomials are orthogonal with respect to the measure $1_{(x>0)} e^{-x} x^\alpha dx$.

The polynomials orthogonal with respect to $\langle .,. \rangle_3$ or equivalently in this case with respect to the measure $x^2 \nu(dx) = 1_{(x>0)} x e^{-x}$ are thus the Laguerre polynomials $L_n^{(1)}(x)$. As mentioned above, the kernel polynomials will be orthogonal with respect to the measure $x^3\nu(dx) = 1_{(x>0)}x^2e^{-x}$. So they are just, up to a constant, the Laguerre polynomials $L_n^{(2)}(x)$. Now a straightforward calculation gives:

$$\begin{aligned} L_n^{1,\sigma^2}(x) &= b_{n,n}x^n + b_{n-1,n}x^{n-1} + \ldots + b_{1,n}x + b_{0,n} \\ &= L_n^{(1)}(x) + \sigma^2 L_n^{(1)}(x)L_{n-1}^{(2)}(0) - \sigma^2 L_n^{(1)}(0)L_{n-1}^{(2)}(x) \\ &= \left(1 + \sigma^2\frac{n(n+1)}{2}\right) L_n^{(1)}(x) - \sigma^2(n+1)L_{n-1}^{(2)}(x). \end{aligned}$$

We can conclude using (5.13) that

$$a_{n,n} = b_{n-1,n-1} = \left(1 + \sigma^2\frac{(n-1)n}{2}\right)\frac{(-1)^{n-1}}{(n-1)!},$$

and that for $k = 1, \ldots, j-1$,

$$\begin{aligned} a_{n,k} &= b_{n-1,k-1} \\ &= \left(1 + \sigma^2\frac{n(n-1)}{2}\right)\frac{(-n+1)_{k-1}(k+1)_{n-k}}{(n-1)!(k-1)!} \\ &\quad - \frac{\sigma^2 n(-n+2)_{k-1}(k+2)_{n-k}}{(n-2)!(k-1)!}. \end{aligned}$$

Notes

The subject of homogeneous chaos, initiated by Wiener [112], has recently received increasing attention [18] [92] and is used in the Malliavin Calculus and Quantum Probability [82].

The stochastic integrals of the Hermite polynomials evaluated in Brownian motion play a fundamental role in the Black–Scholes option pricing model [21]. The binomial process is an important model for the binary market [41], where the prices of a risky asset, namely, the stock, are supposed to jump from one value to one of two possible values at every trading time. Whether the stochastic integrals (or sums) of the Krawtchouk polynomials play a similar role in this theory is the subject of further investigation.

Up to now there have been three stochastic integral relations between stochastic processes and orthogonal polynomials: Hermite–Brownian motion, the Charlier–Poisson process, and the Krawtchouk-binomial process. One could look for other similar stochastic integral relations. Some candidates are the Gamma process (which perhaps can be related to Laguerre polynomials), the Pascal process (Meixner polynomials?), and the Meixner process (Meixner–Pollaczek polynomials?). In [93] however, it was proven that the Brownian motion and the Poisson process are the only normal martingales *with* the classical CRP that possess similar properties. Using the above-derived CRP for Lévy processes, one could look for new similar relations.

Another direction for further investigation can be the application of the new PRP for the hedging of options. In the PRP for Brownian motion (the Black–Scholes model [21]), the predictable ϕ_s is related to the dynamic strategy one has to follow for hedging some derivative security. Unfortunately, Brownian motion is a poor model for stock returns and other Lévy processes where proposed [43]. The new PRP gives a sequence of predictable processes $\{\phi_s^{(i)}, i = 1, 2, \ldots\}$, which perhaps can be used in a similar way. Also possible "approximations" of hedging strategies, using, for example, only $\phi_s^{(1)}$, can be the subject of further investigation. This approximation can possibly be justified by the fact that powers of very small jumps become even smaller and can be ignored; one has to be careful though with the big jumps that our Lévy process makes.

6

Stein Approximation and Orthogonal Polynomials

Stein's method provides a way of finding approximations to the distribution of a random variable, which at the same time gives estimates of the approximation error involved. The strengths of the method are that it can be applied in many circumstances in which dependence plays a part. A key tool in Stein's theory is the generator method developed by Barbour [10]. In this chapter we show how orthogonal polynomials appear in this context in a natural way. They are used in spectral representations of the transition probabilities of Barbour's Markov processes. A good introduction to Stein's approximation method can be found in [95].

6.1 Stein's Method

6.1.1 Standard Normal Distribution

In 1972, Stein [108] published a method to prove normal approximation. It is based on the fact that a random variable Z has a standard normal distribution $\mathrm{N}(0,1)$ if and only if for all differentiable functions f such that $E|f'(X)| < \infty$, where X has a standard normal distribution $\mathrm{N}(0,1)$,

$$E[f'(Z) - Zf(Z)] = 0. \tag{6.1}$$

Stein Equation

Hence, it seems reasonable that if $E[f'(W) - Wf(W)]$ is small for a large class of functions f, then the distribution of W is close to the standard normal distribution. Suppose we wish to estimate the difference between the expectation of a smooth function h with respect to the random variable W and $E[h(Z)]$, where Z has a standard normal distribution. Stein [108] showed that for any smooth, real-valued bounded function h there is a function $f = f_h$ solving the now-called *Stein equation* for the standard normal distribution

$$f'(x) - xf(x) = h(x) - E[h(Z)], \tag{6.2}$$

with Z a standard normal random variable. The unique bounded solution of the above equation is given by

$$f_h(x) = \exp(x^2/2) \int_{-\infty}^{x} (h(y) - E[h(Z)]) \exp(-y^2/2) dy.$$

Then we estimate

$$E[f_h'(W) - Wf_h(W)] \tag{6.3}$$

and hence $E[h(W)] - E[h(Z)]$. The next step is to show that the quantity (6.3) is small. In order to do this we use the structure of W. For instance, it might be that W is a sum of independent random variables. In addition we use some smoothness conditions on f_h. Stein showed the following inequalities

$$\begin{aligned} ||f_h|| &\leq \sqrt{\frac{\pi}{2}} ||h - E[h(Z)]||, & (6.4) \\ ||f_h'|| &\leq \sup(h) - \inf(h), \\ ||f_h''|| &\leq 2||h'||, & (6.5) \end{aligned}$$

where $|| \cdot ||$ denotes the supremum norm.

In this way we can bound the distance of W from the normal, in terms of a test function h; the immediate bound of the distance is one of the key advantages of Stein's method compared to moment-generating functions or characteristic functions.

Central Limit Theorem

To see how Stein's method works, consider a sequence of independent and identically distributed random variables, $X_1, X_2, \ldots, X_n$ with common zero mean, unit variance, and $\xi = E[|X_i|^3] < \infty$, and let W_n be its normalized sum,

$$W_n = \frac{\sum_{i=1}^{n} X_i}{\sqrt{n}}.$$

Let $h : \mathbb{R} \to \mathbb{R}$ be a continuously differentiable function such that $||h'|| < \infty$. Let W_n^i be the normalized sum of the X_j without the ith term; i.e.,

$W_n^i = W_n - X_i/\sqrt{n}$. By using Taylor's formula with remainder, expanding $f_h(W_n)$ about W_n^i, we obtain

$$f_h(W_n) - \big(f_h(W_n^i) + (W_n - W_n^i)f_h'(W_n^i)\big) = \int_{W_n^i}^{W_n} (W_n - t)df_h'(t).$$

Multiplying both sides by $X_i/\sqrt{n}$ and taking expectation, we have

$$\begin{aligned} & E\left[\frac{X_i}{\sqrt{n}}f_h(W_n) - \frac{X_i}{\sqrt{n}}f_h(W_n^i) - \frac{X_i^2}{n}f_h'(W_n^i)\right] \\ &= E\left[\frac{X_i}{\sqrt{n}}\int_{W_n^i}^{W_n}(W_n - t)df_h'(t)\right]. \end{aligned}$$

Now taking absolute value on both sides, we may bound the right-hand side by

$$E\left[\left|\frac{X_i}{\sqrt{n}}\int_{W_n^i}^{W_n}||f_h''|| \quad |W_n - W_n^i|dt\right|\right] = ||f_h''||E[|X_i|^3]n^{-3/2}.$$

Also, since X_i and W_n^i are independent,

$$\begin{aligned} E[X_i f_h(W_n^i)] &= E[X_i]E[f_h(W_n^i)] = 0 \\ E[X_i^2 f_h'(W_n^i)] &= E[X_i^2]E[f_h'(W_n^i)] = E[f_h'(W_n^i)]. \end{aligned}$$

Therefore we have

$$\left|E\left[\frac{X_i}{\sqrt{n}}f(W_n)\right] - \frac{1}{n}E[f'(W_n^i)]\right| \leq ||f_h''||\xi n^{-3/2}. \tag{6.6}$$

Now we may estimate $|E[h(W_n)] - E[h(Z)]|$ by

$$\begin{aligned} & |E[f_h'(W_n) - W_n f(W_n)]| \\ \leq & \left|E\left[f_h'(W_n) - \frac{\sum_{i=1}^n f_h'(W_n^i)}{n}\right]\right| \\ & + \left|E\left[\frac{\sum_{i=1}^n f_h'(W_n^i)}{n} - \frac{\sum_{i=1}^n X_i f(W_n)}{\sqrt{n}}\right]\right| \\ \leq & \frac{\sum_{i=1}^n E|f_h'(W_n) - f_h'(W_n^i)|}{n} \\ & + \sum_{i=1}^n \left|\frac{E[f_h'(W_n^i)]}{n} - \frac{E[X_i f(W_n)]}{\sqrt{n}}\right|. \end{aligned} \tag{6.7}$$

Applying the mean value theorem and Jensen's inequality, we obtain the following bound for the first term in the above expression

$$\frac{1}{n}\sum_{i=1}^n ||f_h''||\frac{E|X_i|}{\sqrt{n}} \leq ||f_h''||\frac{\xi}{\sqrt{n}}.$$

By (6.6), the second term of (6.7) can be bounded by

$$||f_h''||\xi/\sqrt{n}.$$

Using the smoothness estimate (6.5), we replace $||f_h''||$ by its bound $2||h'||$. Hence, we obtain the Berry–Esseen type bound

$$|E[h(W_n)] - E[h(Z)]| \leq 4||h'||\frac{\xi}{\sqrt{n}}.$$

6.1.2 *Poisson Distribution*

Chen [23] applied Stein's idea in the context of Poisson approximation. The Stein equation for the Poisson distribution $\mathrm{P}(\mu)$ is now a difference equation:

$$\mu f(x) - xf(x-1) = h(x) - E[h(Z)], \tag{6.8}$$

where h is a bounded real-valued function defined on the set of the nonnegative integers and Z has a Poisson distribution $\mathrm{P}(\mu)$. The choice of the left-hand side of Equation (6.8) is based on the fact that a random variable W on a set of nonnegative integers has a Poisson distribution with parameter μ if and only if for all bounded real-valued functions f on the integers

$$E[\mu f(W) - Wf(W-1)] = 0.$$

The solution of the Stein equation (6.8) for the Poisson distribution $\mathrm{P}(\mu)$ is given by:

$$f_h(x) = x!\mu^{-x-1}\sum_{k=0}^{x}(h(k) - E[h(Z)])\frac{\mu^k}{k!}, \quad x \geq 0. \tag{6.9}$$

This solution is the unique, except at $x < 0$, bounded solution; the value $f_h(x)$ for negative x does not enter into consideration and is conventionally taken to be zero. In [11] one finds the following estimates of the smoothness for f_h by an analytic argument.

$$\begin{aligned} ||f_h|| &\leq ||h||\min(1, \mu^{-1/2}), \\ ||\Delta f_h|| &\leq ||h||\min(1, \mu^{-1}), \end{aligned}$$

where $\Delta f(x) = f(x+1) - f(x)$.

6.1.3 *General Procedure*

For an arbitrary distribution ρ, the general procedure is: find a good characterization of the desired distribution ρ in terms of an equation, that is, of the type

$$Z \text{ is a r.v. with distribution } \rho \text{ if and only if } E[\mathcal{A}f(Z)] = 0,$$

for all smooth functions f, where A is an operator associated with the distribution ρ. (Thus in the standard normal case $Af(x) = f'(x) - xf(x)$, $x \in \mathbb{R}$.) We call such an operator a *Stein operator*. Next assume Z to have distribution ρ, and consider the Stein equation

$$h(x) - E[h(Z)] = Af(x). \tag{6.10}$$

For every smooth h, find a corresponding solution f_h of this equation. For any random variable W,

$$E[h(W)] - E[h(Z)] = E[Af_h(W)].$$

Hence, to estimate the proximity of W and Z, it is sufficient to estimate $E[Af_h(W)]$ for all possible solutions of (6.10).

However, in this procedure it is not completely clear which characterizing equation for the distribution to choose (one could think of a whole set of possible equations). The aim is to be able to solve (6.10) for a sufficiently large class of functions h, to obtain convergence in a known topology.

6.2 The Generator Method

A key tool in Stein's theory is the generator method developed by Barbour [10]. Replacing f by f' in the Stein equation for the standard normal (6.2) gives

$$f''(x) - xf'(x) = h(x) - E[h(Z)]. \tag{6.11}$$

If we set $\mathcal{A}f(x) = f''(x) - xf'(x)$, this equation can be rewritten as

$$\mathcal{A}f = h(x) - E[h(Z)].$$

The key advantage is that $\mathcal{A}$ is also the generator of a Markov process, the Ornstein–Uhlenbeck process, with standard normal stationary distribution.

If we replace f by $\Delta f = f(x+1) - f(x)$ in the Stein equation for the Poisson distribution (6.8), we get

$$\mu f(x+1) - (\mu + x)f(x) + xf(x-1) = h(x) - E[h(Z)]. \tag{6.12}$$

If we set $\mathcal{A}f(x) = \mu f(x+1) - (\mu + x)f(x) + xf(x-1)$, this equation can be rewritten as

$$\mathcal{A}f(x) = h(x) - E[h(Z)].$$

Again we see that $\mathcal{A}$ is a generator of a Markov process, an immigration–death process, with stationary distribution the Poisson distribution. Indeed from (2.5) we see that $\mathcal{A}$ is the generator of a birth and death process with constant birth (or immigration) rate $\lambda_i = \mu$ and linear death rate $\mu_i = i$.

Stein–Markov Operator and Stein–Markov Equation

This also works for a broad class of other distributions ρ. Barbour suggested employing for an operator the generator of a Markov process. So for a random variable Z with distribution ρ, we are looking for an operator $\mathcal{A}$, such that $E[\mathcal{A}f(Z)] = 0$ and for a Markov process $\{X_t, t \geq 0\}$ with generator $\mathcal{A}$ and with unique stationary distribution ρ. We call such an operator $\mathcal{A}$ a *Stein–Markov operator* for ρ. The associated equation is called the *Stein–Markov equation*

$$\mathcal{A}f(x) = h(x) - E[h(Z)]. \tag{6.13}$$

In the following this method is called *the generator method.*

However, for a given distribution ρ, there may be various operators $\mathcal{A}$ and Markov processes with ρ as stationary distributions. We provide a general procedure to obtain one such process for a large class of distributions.

In this framework, for a bounded function h, the solution to the Stein–Markov equation (6.13) may be given by

$$f_h(x) = -\int_0^\infty (T_t h(x) - E[h(Z)])dt, \tag{6.14}$$

where Z has distribution ρ, X_t is a Markov process with generator $\mathcal{A}$ and stationary distribution ρ, and

$$T_t h(x) = E[h(X_t)|X_0 = x]. \tag{6.15}$$

6.3 Stein Operators

We summarize some Stein–Markov operators $\mathcal{A}$ and Stein operators A for some well-known distributions in Tables 6.1 and 6.2. For more details see [8], [9], [12], [23], [77], [108], and references cited therein. For convenience we put $q = 1 - p$.

TABLE 6.1. Stein Operators

Name	**Notation**	$Af(x)$
Normal	$\mathrm{N}(0,1)$	$f''(x) - xf'(x)$
Poisson	$\mathrm{P}(\mu)$	$\mu f(x+1) - (x+\mu)f(x) + xf(x-1)$
Gamma	$\mathrm{G}(a,1)$	$xf''(x) + (a+1-x)f'(x)$
Pascal	$\mathrm{Pa}(\gamma,\mu)$	$\mu(x+\gamma)f(x+1) - (\mu(x+\gamma)+x)f(x)$ $+xf(x-1)$
Binomial	$\mathrm{Bin}(N,p)$	$p(N-x)f(x+1) - (p(N-x)+qx)f(x)$ $+qxf(x-1)$

TABLE 6.2. Stein–Markov Operators

Name	**Notation**	$\mathcal{A}f(x)$
Normal	$\mathrm{N}(0,1)$	$f'(x) - xf(x)$
Poisson	$\mathrm{P}(\mu)$	$\mu f(x+1) - xf(x)$
Gamma	$\mathrm{G}(a,1)$	$xf'(x) + (a+1-x)f(x)$
Pascal	$\mathrm{Pa}(\gamma,\mu)$	$\mu(x+\gamma)f(x) - xf(x-1)$
Binomial	$\mathrm{Bin}(N,p)$	$p(N-x)f(x) - qxf(x-1)$

Note that the Stein–Markov and Stein operators are of the form

$$\begin{aligned} \mathcal{A}f(x) &= s(x)f''(x) + \tau(x)f'(x) \\ Af(x) &= s(x)f'(x) + \tau(x)f(x) \end{aligned}$$

in the continuous case and of the form

$$\begin{aligned} &\mathcal{A}f(x) \\ &\quad = s(x)\Delta\nabla f(x) + \tau(x)\Delta f(x) \\ &\quad = (s(x)+\tau(x))f(x+1) - (2s(x)+\tau(x))f(x) + s(x)f(x-1) \\ &Af(x) \\ &\quad = s(x)\nabla f(x) + \tau(x)f(x) \\ &\quad = (s(x)+\tau(x))f(x) - s(x)f(x-1) \end{aligned}$$

in the discrete case, where the $s(x)$ and $\tau(x)$ are polynomials of degree at most two and one, respectively. Furthermore the above distributions (see Chapter 1) satisfy equations with the same ingredients $s(x)$ and $\tau(x)$. In the continuous case the density (or weight) function $\rho(x)$ of the distribution ρ satisfies the differential equation (1.7),

$$(s(x)\rho(x))' = \tau(x)\rho(x),$$

and in the discrete case the probabilities $\Pr(Z=x) = p_x$ satisfy the difference equation (1.11),

$$\Delta(s(x)p_x) = \tau(x)p_x.$$

This brings us to the Pearson class of continuous distributions and Ord's family of discrete distributions.

6.4 Stein's Method for Pearson and Ord Families

6.4.1 The Pearson Family of Continuous Distributions

In 1895 K. Pearson introduced his famous family of frequency curves. The elements of this family arise by considering the possible solutions to the differential equation

$$\rho'(x) = \frac{(x+a_0)\rho(x)}{b_0 + b_1 x + b_2 x^2} = \frac{q(x)\rho(x)}{p(x)}. \tag{6.16}$$

There are in essence five basic solutions, depending on whether the polynomial $p(x)$ in the denominator is constant, linear or quadratic and, in the last case, on whether the discriminant, $D = b_1^2 - 4b_0b_2$, of $p(x)$ is positive, negative, or zero. It is easy to show that the Pearson family is closed under translation and scale change. Thus the study of the family can be reduced to differential equations that result after an affine transformation of the independent variable.

1. If $\deg(p(x)) = 0$, then $\rho(x)$ can be reduced after a change of variable to a standard normal density.

2. If $\deg(p(x)) = 1$, then the resulting solution may be seen to be the family of Gamma distributions.

3. If $\deg(p(x)) = 2$ and $D = 0$, then the density $\rho(x)$ is of the form

   $$\rho(x) = Cx^{-\alpha} \exp(-\beta/x),$$

   where C is the appropriate normalizing constant.

4. If $\deg(p(x)) = 2$ and $D < 0$, then the density $\rho(x)$ can be brought into the form

   $$\rho(x) = C(1+x^2)^{-\alpha} \exp(\beta \arctan(x)),$$

   where again C is the appropriate normalizing constant; in particular, the t-distributions are a rescaled subfamily of this class.

5. If $\deg(p(x)) = 2$ and $D > 0$, the density $\rho(x)$ can be brought into the form

   $$\rho(x) = Cx^{\alpha-1}(1-x)^{\beta-1},$$

   where C is the appropriate normalizing constant; the Beta densities clearly belong to this class.

In what follows we suppose that in the continuous case we have a distribution ρ on an interval (a, b), with a and b possibly infinite, with a second moment, a distribution function $F(x)$, and a density function $\rho(x)$, but we find it more convenient to work with an equivalent form of the differential equation (6.16). We assume that our density function $\rho(x)$ satisfies:

$$(s(x)\rho(x))' = \tau(x)\rho(x), \tag{6.17}$$

for some polynomials $s(x)$ of degree at most two and $\tau(x)$ of exact degree one. The equivalence between (6.16) and (6.17) can easily be seen by setting $p(x) = s(x)$ and $q(x) = \tau(x) - s'(x)$.

Furthermore we make the following assumptions on $s(x)$.

$$s(x) > 0, \quad a < x < b \text{ and } s(a), s(b) = 0 \text{ if } a, b \text{ is finite.} \tag{6.18}$$

Note that because $\rho(x) \geq 0$ and $\int_a^b \rho(y)dy = 1$, we have that $\tau(x)$ is not a constant and is a decreasing linear function. Indeed, suppose it were nonconstant and increasing and denote the only zero of $\tau(x)$ by l; then we would have for $x < l$,

$$\int_a^x \rho(y)dy \leq \int_a^x \frac{\tau(y)}{\tau(x)}\rho(y)dy = \frac{\int_a^x (s(y)\rho(y))'dy}{\tau(x)} = \frac{s(x)\rho(x)}{\tau(x)} < 0,$$

which is impossible. For a similar reason, $\tau(x)$ cannot be constant.

The only zero of $\tau(x)$, l say, is just $E[Z]$, where Z has distribution ρ. This can be seen by calculating,

$$E[\tau(Z)] = \int_a^b \tau(y)\rho(y)dy = \int_a^b (s(y)\rho(y))'dy = s(y)\rho(y)|_a^b = 0.$$

We start with a characterization of a distribution ρ with density $\rho(x)$ satisfying (6.17). We set C_1 equal to the set of all real bounded piecewise continuous functions on the interval (a, b) and set C_2 equal to the set of all real continuous and piecewise continuously differentiable functions f on the interval (a, b), for which the function $g(z) \equiv |s(z)f'(z)| + |\tau(z)f(z)|$ is bounded. We have the following theorem.

Theorem 13 *Suppose we have a random variable X on (a, b) with density function $\tilde{\rho}(x)$ and finite second moment, and that $\rho(x)$ satisfies (6.17). Then $\tilde{\rho}(x) = \rho(x)$ if and only if for all functions $f \in C_2$,*

$$E[s(X)f'(X) + \tau(X)f(X)] = 0.$$

Proof: First assume X has density function $\rho(x)$. Then

$$\begin{aligned}
&E[s(X)f'(X) + \tau(X)f(X)] \\
&= \int_a^b (s(x)f'(x) + \tau(x)f(x))\rho(x)dx \\
&= \int_a^b f'(x)(s(x)\rho(x))dx + \int_a^b f(x)\tau(x)\rho(x)dx \\
&= f(x)s(x)\rho(x)|_a^b - \int_a^b f(x)(s(x)\rho(x))'dx \\
&\quad + \int_a^b f(x)\tau(x)\rho(x)dx \\
&= -\int_a^b f(x)\tau(x)\rho(x)dx + \int_a^b f(x)\tau(x)\rho(x)dx \\
&= 0.
\end{aligned}$$

Conversely, suppose we have a random variable X on (a, b) with density function $\tilde{\rho}(x)$ and finite second moment such that for all functions $f \in \mathcal{C}_2$,

$$E[s(X)f'(X) + \tau(X)f(X)] = 0.$$

Then

$$\begin{aligned} 0 &= E[s(X)f'(X) + \tau(X)f(X)] \\ &= \int_a^b (s(x)f'(x) + \tau(x)f(x))\tilde{\rho}(x)dx \\ &= f(x)\tilde{\rho}(x)s(x)|_a^b - \int_a^b (s(x)\tilde{\rho}(x))'f(x)dx + \int_a^b \tau(x)f(x)\tilde{\rho}(x)dx \\ &= -\int_a^b (s(x)\tilde{\rho}(x))'f(x)dx + \int_a^b \tau(x)f(x)\tilde{\rho}(x)dx. \end{aligned}$$

But this means that for all functions $f \in \mathcal{C}_2$,

$$\int_a^b (s(x)\tilde{\rho}(x))'f(x)dx = \int_a^b (\tau(x)\tilde{\rho}(x))f(x)dx.$$

So $\tilde{\rho}(x)$ satisfies the differential equation

$$(s(x)\tilde{\rho}(x))' = \tau(x)\rho(x),$$

which uniquely defines the density $\rho(x)$. In conclusion we have $\tilde{\rho}(x) = \rho(x)$. ◇

In Stein's method we wish to estimate the difference between the expectation of a function $h \in \mathcal{C}_1$ with respect to a random variable W and $E[h(Z)]$, where Z has distribution ρ. To do this, we first solve the so-called Stein equation for the distribution ρ,

$$s(x)f'(x) + \tau(x)f(x) = h(x) - E[h(Z)]. \tag{6.19}$$

The solution of this Stein equation is given in the next proposition.

Proposition 4 *The Stein equation (6.19) for the distribution ρ and a function $h \in \mathcal{C}_1$ has as solution*

$$\begin{aligned} f_h(x) &= \frac{1}{s(x)\rho(x)} \int_a^x (h(y) - E[h(Z)])\rho(y)dy && (6.20) \\ &= \frac{-1}{s(x)\rho(x)} \int_x^b (h(y) - E[h(Z)])\rho(y)dy, && (6.21) \end{aligned}$$

when $a < x < b$ and $f_h = 0$ elsewhere. This f_h belongs to $\mathcal{C}_2$.

Proof: First note that

$$
\begin{aligned}
f_h'(x) &= \frac{-(s(x)\rho(x))'}{(s(x)\rho(x))^2}\int_a^x (h(y)-E[h(Z)])\rho(y)dy + \frac{h(x)-E[h(Z)]}{s(x)} \\
&= \frac{-\tau(x)\rho(x)}{(s(x)\rho(x))^2}\int_a^x (h(y)-E[h(Z)])\rho(y)dy + \frac{h(x)-E[h(Z)]}{s(x)} \\
&= \frac{-\tau(x)}{(s(x))^2\rho(x)}\int_a^x (h(y)-E[h(Z)])\rho(y)dy + \frac{h(x)-E[h(Z)]}{s(x)}.
\end{aligned}
$$

Next we just substitute the proposed solution (6.20) into the left-hand side of the Stein equation. This gives

$$
\begin{aligned}
& s(x)f_h'(x) + \tau(x)f_h(x) \\
&= \frac{-\tau(x)}{s(x)\rho(x)}\int_a^x (h(y)-E[h(Z)])\rho(y)dy + h(x) - E[h(Z)] \\
&\quad + \frac{\tau(x)}{s(x)\rho(x)}\int_a^x (h(y)-E[h(Z)])\rho(y)dy \\
&= h(x) - E[h(Z)].
\end{aligned}
$$

The second expression for f_h follows from the fact

$$
\int_a^x (h(y)-E[h(Z)])\rho(y)dy + \int_x^b (h(y)-E[h(Z)])\rho(y)dy = 0.
$$

To prove that for $h \in C_1$ we have $f \in C_2$, we need only show that $g(x) \equiv |s(x)f_h'(x)| + |\tau(x)f_h(x)|$, $a < x < b$, is bounded. We have for $x < l$,

$$
\begin{aligned}
g(x) &= |s(x)f_h'(x)| + |\tau(x)f_h(x)| \\
&\le \left|\frac{\tau(x)}{s(x)\rho(x)}\int_a^x (h(y)-E[h(Z)])\rho(y)dy + h(x) - E[h(Z)]\right| \\
&\quad + \left|\frac{\tau(x)}{s(x)\rho(x)}\int_a^x (h(y)-E[h(Z)])\rho(y)dy\right| \\
&\le \left|\frac{||h(x)-E[h(Z)]||}{s(x)\rho(x)}\int_a^x \tau(y)\rho(y)dy\right| + ||h(x)-E[h(Z)]|| \\
&\quad + \left|\frac{||h(y)-E[h(Z)]||}{s(x)\rho(x)}\int_a^x \tau(y)\rho(y)dy\right| \\
&\le 3||h(x)-E[h(Z)]||,
\end{aligned}
$$

where $||f(x)|| = \sup_{a<x<b}|f(x)|$. A similar result for $x \ge l$ follows from (6.21). This proves our proposition. ◇

The next step is to estimate

$$
E[s(W)f_h'(W) + \tau(W)f_h(W)] \tag{6.22}
$$

and hence $E[h(W)] - E[h(Z)]$. To show the quantity in (6.22) is small, it is necessary to use the structure of W. In addition we might require certain smoothness conditions on f_h that would translate into smoothness conditions on h by the following lemma.

Set as before $l = E[Z]$ and remember that l is the only zero of τ. We need the following positive constant

$$M = \frac{1}{\rho(l)s(l)} \max(F(l), 1 - F(l)).$$

Lemma 2 *Suppose $h \in C_1$ and let f_h be the solution (6.20) of the Stein equation given by (6.19). Then*

$$||f_h(x)|| \leq M||h(x) - E[h(Z)]||. \tag{6.23}$$

Proof: Note that $\tau(x)$ is positive and decreasing in (a, l) and negative and decreasing in (l, b).

So we have for $x < l$,

$$\begin{aligned} F(x) = \int_a^x \rho(y)dy &\leq \int_a^x \frac{\tau(y)}{\tau(x)} \rho(y)dy \\ &= \frac{1}{\tau(x)} \int_a^x (s(y)\rho(y))'dy \\ &= \frac{s(x)\rho(x)}{\tau(x)}. \end{aligned} \tag{6.24}$$

Similarly, for $x > l$, we have

$$1 - F(x) = \int_x^b \rho(y)dy \leq \frac{-s(x)\rho(x)}{\tau(x)}. \tag{6.25}$$

Now, for $x \leq l$,

$$\begin{aligned} |f_h(x)| &= \left| \frac{1}{s(x)\rho(x)} \int_a^x (h(y) - E[h(Z)])\rho(y)dy \right| \\ &\leq ||h(x) - E[h(Z)]|| \frac{\int_a^x \rho(y)dy}{s(x)\rho(x)}. \end{aligned} \tag{6.26}$$

Similarly for $x \geq l$,

$$\begin{aligned} |f_h(x)| &= \left| \frac{-1}{s(x)\rho(x)} \int_x^b (h(y) - E[h(Z)])\rho(y)dy \right| \\ &\leq ||h(x) - E[h(Z)]|| \frac{\int_x^b \rho(y)dy}{s(x)\rho(x)}. \end{aligned} \tag{6.27}$$

Next we prove that the expressions

$$\frac{\int_a^x \rho(y)dy}{s(x)\rho(x)} \text{ and } \frac{\int_x^b \rho(y)dy}{s(x)\rho(x)}$$

of (6.26) and (6.27) attain their maximum at $x = l$. To show this we calculate

$$\left(\frac{\int_a^x \rho(y)dy}{s(x)\rho(x)}\right)' = \frac{-\tau(x)}{\rho(x)(s(x))^2}\int_a^x \rho(y)dy + \frac{1}{s(x)}$$

and

$$\left(\frac{\int_x^b \rho(y)dy}{s(x)\rho(x)}\right)' = \frac{-\tau(x)}{\rho(x)(s(x))^2}\int_x^b \rho(y)dy - \frac{1}{s(x)}.$$

Next we use (6.24) and (6.25) to obtain for $x \leq l$,

$$\left(\frac{\int_a^x \rho(y)dy}{s(x)\rho(x)}\right)' \geq 0$$

and for $x \geq l$,

$$\left(\frac{\int_x^b \rho(y)dy}{s(x)\rho(x)}\right)' \leq 0.$$

In conclusion we have $||f_h(x)|| \leq M||h - E[h(Z)]||$. ◇

Lemma 3 *Suppose $h \in C_1$ and let f_h be the solution (6.20) of the Stein equation given by (6.19). Then*

$$||f_h'(x)|| \leq 2||1/s(x)|| \times ||h(x) - E[h(Z)]||. \tag{6.28}$$

Proof: Because

$$\begin{aligned} f_h'(x) &= \frac{-\tau(x)}{(s(x))^2\rho(x)}\int_a^x (h(y) - E[h(Z)])\rho(y)dy + \frac{h(x) - E[h(Z)]}{s(x)} \\ &= \frac{\tau(x)}{(s(x))^2\rho(x)}\int_x^b (h(y) - E[h(Z)])\rho(y)dy + \frac{h(x) - E[h(Z)]}{s(x)}, \end{aligned}$$

we have for $x \leq l$,

$$\begin{aligned} |f_h'(x)| &\leq ||h(x) - E[h(Z)]|| \left(\frac{\tau(x)F(x)}{(s(x))^2\rho(x)} + \frac{1}{s(x)}\right) \\ &\leq ||h(x) - E[h(Z)]|| \left(\frac{2}{s(x)}\right), \end{aligned}$$

where we used (6.24) for the last inequality.

Similarly, for $x \geq l$ we have

$$\begin{aligned} |f_h'(x)| &\leq ||h(x) - E[h(Z)]|| \left(\frac{-\tau(x)(1 - F(x))}{(s(x))^2\rho(x)} + \frac{1}{s(x)}\right) \\ &\leq ||h(x) - E[h(Z)]|| \left(\frac{2}{s(x)}\right). \end{aligned}$$

This ends the proof. ◇

Examples

1. The normal distribution $N(m, \sigma^2)$ with mean $m \in \mathbb{R}$ and variance $\sigma^2 > 0$, has a density function

$$\rho(x; m, \sigma^2) = \frac{\exp(-(x-m)^2/(2\sigma^2))}{\sqrt{2\pi\sigma^2}}, \quad x \in \mathbb{R}.$$

Clearly we have

$$\frac{\rho'(x; m, \sigma^2)}{\rho(x; m, \sigma^2)} = \frac{m-x}{x} = \frac{q(x)}{p(x)},$$

and thus $s(x) = p(x) = \sigma^2$ and $\tau(x) = q(x) + s'(x) = m - x$.
So the Stein equation for the $N(m, \sigma^2)$ distribution is given by

$$\sigma^2 f'(x) + (m-x)f(x) = h(x) - E[h(Z)].$$

The Stein operator is given by

$$Af(x) = \sigma^2 f'(x) + (m-x)f(x)$$

and

$$M = \sqrt{\frac{\pi}{2\sigma^2}}.$$

So for the standard normal distribution $N(0,1)$, Lemma 2 recovers the first bound in (6.4).
This case was the starting point of Stein's theory [108].

2. The Gamma distribution $G(r, \lambda^{-1})$, has a density function

$$\rho(x; r, \lambda) = \frac{\lambda^r}{\Gamma(r)} e^{-\lambda x} x^{r-1}, \quad x > 0.$$

Clearly we have

$$\frac{\rho'(x; r, \lambda)}{\rho(x; r, \lambda)} = \frac{(r-1) - \lambda x}{x} = \frac{q(x)}{p(x)},$$

and thus $s(x) = p(x) = x$ and $\tau(x) = q(x) + s'(x) = r - \lambda x$.
So the Stein equation for the $G(r, \lambda^{-1})$ distribution is given by

$$x f'(x) + (r - \lambda x) f(x) = h(x) - E[h(Z)].$$

This case is treated by H.M. Luk [77].

3. The Beta distribution $\mathrm{B}(\alpha, \beta)$ on $(0,1)$, with parameters $\alpha, \beta > 0$, has a density function

$$\rho(x; \alpha, \beta) = \frac{x^{\alpha-1}(1-x)^{\beta-1}}{B(\alpha, \beta)}, \quad 0 < x < 1.$$

Clearly we have

$$\frac{\rho'(x; \alpha, \beta)}{\rho(x; \alpha, \beta)} = \frac{(-\alpha - \beta + 2)x + \alpha - 1}{(1-x)x} = \frac{q(x)}{p(x)},$$

and thus $s(x) = p(x) = (1-x)x$ and $\tau(x) = q(x) + s'(x) = -(\alpha + \beta)x + \alpha$.

So the Stein equation for the $\mathrm{B}(\alpha, \beta)$ distribution is given by

$$x(1-x)f'(x) + (\alpha - (\alpha + \beta)x)f(x) = h(x) - E[h(Z)].$$

This Stein equation seems to be new.

4. The Student's t-distribution t_n with $n \in \{1, 2, \ldots\}$ degrees of freedom has a density function

$$\rho(x; n) = \frac{\Gamma((n+1)/2)}{\sqrt{n\pi}\Gamma(n/2)} \left(1 + \frac{x^2}{n}\right)^{-(n+1)/2}, \quad x \in \mathbb{R}.$$

Clearly we have

$$\frac{\rho'(x; n)}{\rho(x; n)} = \frac{-(n+1)x/n}{1 + x^2/n} = \frac{q(x)}{p(x)},$$

and thus clearly $s(x) = p(x) = 1 + (x^2/n)$ and $\tau(x) = q(x) + s'(x) = -((n-1)/n)x$.

So the Stein equation for the t_n distribution is given by

$$\left(1 + \frac{x^2}{n}\right) f'(x) - \frac{n-1}{n} x f(x) = h(x) - E[h(Z)].$$

This Stein equation seems to be new. Note that Lemma 3 gives us a useful bound on f_h', namely,

$$||f_h'|| \leq 2||h - E[h(Z)]||.$$

6.4.2 Ord's Family of Discrete Distributions

Ord's family comprises all the discrete distributions that satisfy

$$\frac{\Delta p_x}{p_x} = \frac{p_{x+1} - p_x}{p_x} = \frac{a_0 + a_1 x}{b_0 + b_1 x + b_2 x^2} = \frac{q(x)}{p(x)}, \tag{6.29}$$

where $p_x = \Pr(Z = x)$ and x takes values in $S = \{a, a+1, \dots, b-1, b\}$, with a, b possibly infinite and where we set for convenience $p_x = 0$ for $x \notin S$.

So we suppose that we have a discrete distribution ρ on S with a finite second moment, but we find it more convenient to work with an equivalent form of the difference equation (6.29). We assume that our probabilities p_x satisfy

$$\Delta(s(x)p_x) = \tau(x)p_x, \tag{6.30}$$

for some polynomials $s(x)$ of degree at most two and $\tau(x)$ of exact degree one. The equivalence between (6.30) and (6.29) can easily be seen by using

$$\Delta(s(x)p_x) = s(x+1)\Delta p_x + p_x \Delta s(x),$$

and setting $p(x) = s(x+1)$ and $q(x) = \tau(x) - \Delta s(x)$.

Furthermore we make the following assumptions on $s(x)$,

$$s(a) = 0 \text{ if } a \text{ is finite}, \quad s(x) > 0, \quad a < x \leq b. \tag{6.31}$$

Note that because $p_x \geq 0$ and $\sum_{i=a}^{b} p_i = 1$, that $\tau(x)$ is not a constant and is a decreasing linear function. Indeed, suppose it were nonconstant and increasing and denote the only zero of $\tau(x)$ by l; then we would have for $x < l$,

$$\sum_{i=a}^{x} p_i \leq \sum_{i=a}^{x} \frac{\tau(i)}{\tau(x)} p_i = \frac{\sum_{i=a}^{x} \Delta(s(i)p_i)}{\tau(x)} = \frac{s(x+1)p(x+1)}{\tau(x)} < 0,$$

which is impossible. For a similar reason, $\tau(x)$ cannot be a constant.

The only zero of $\tau(x)$, l say, is just $E[Z]$, where Z has distribution ρ. This can be seen by calculating,

$$E[\tau(Z)] = \sum_{i=a}^{b} \tau(i)p_i = \sum_{i=a}^{b} \Delta(s(i)p_i) = s(i)p_i|_a^{b+1} = 0.$$

We start with a characterization of a distribution ρ with probabilities p_x satisfying (6.30) and (6.31). We set C_3 equal to the set of all real-valued functions f on the integers such that f is zero outside S, the function $g(x) \equiv |s(x)\nabla f(x)| + |\tau(x)f(x)|$ is bounded, and where $\nabla f(x) = f(x) - f(x-1)$. We have the following theorem.

Theorem 14 *Suppose we have a discrete random variable X on the set S with probabilities $\Pr(X = x) = \tilde{p}_x$ and finite second moment and that p_x satisfies (6.30). Then $\tilde{p}_x = p_x$ if and only if for all functions $f \in C_3$,*

$$E[s(X)\nabla f(X) + \tau(X)f(X)] = 0.$$

Proof: First assume X has probabilities p_x. Then

$$\begin{aligned} &E[s(X)\nabla f(X) + \tau(X)f(X)] \\ &= \sum_{i=a}^{b}(s(i)\nabla f(i) + \tau(i)f(i))p_i \\ &= \sum_{i=a}^{b}\nabla f(i)(s(i)p_i) + \sum_{i=a}^{b} f(i)\tau(i)p_i \\ &= f(i)p_i s(i)|_a^{b+1} - \sum_{i=a}^{b} f(i)\Delta(s(i)p_i) \\ &\quad + \sum_{i=a}^{b} f(i)\tau(i)p_i \\ &= -\sum_{i=a}^{b} f(i)\tau(i)p_i + \sum_{i=a}^{b} f(i)\tau(i)p_i \\ &= 0. \end{aligned}$$

Conversely, suppose we have a random variable X on S with probabilities $\tilde{p}_x$, such that for all functions $f \in \mathcal{C}_3$,

$$E[s(X)\nabla f(X) + \tau(X)f(X)] = 0.$$

Then

$$\begin{aligned} 0 &= E[s(X)\nabla f(X) + \tau(X)f(X)] \\ &= \sum_{i=a}^{b}(s(x)\nabla f(x) + \tau(x)f(x))\tilde{p}_i \\ &= -\sum_{i=a}^{b}\Delta(s(i)\tilde{p}_i)f(i) + \sum_{i=a}^{b}\tau(i)f(i)\tilde{p}_i. \end{aligned}$$

But this means that for all functions $f \in \mathcal{C}_3$,

$$\sum_{i=a}^{b}\Delta(s(i)\tilde{p}_i)f(i) = \sum_{i=a}^{b}(\tau(i)\tilde{p}_i)f(i).$$

In particular for f an indicator function of a singleton, i.e., for $f(x) = 1_{(x=j)}$, for all $j \in S$, we have

$$\Delta(s(j)\tilde{p}_j) = (\tau(j)\tilde{p}_j).$$

So $\tilde{p}_x$ satisfies the difference equation (6.30) which uniquely defines the probabilities p_x. In conclusion we have $\tilde{p}_x = p_x$. $\diamond$

As explained before, we wish to estimate the difference between the expectation of a bounded function h with respect to a random variable W and $E[h(Z)]$, where Z has distribution ρ. To do this, we first solve the so-called Stein equation for the distribution ρ,

$$s(x)\nabla f(x) + \tau(x)f(x) = h(x) - E[h(Z)]. \tag{6.32}$$

This Stein equation is solved in the next proposition.

Proposition 5 *The Stein equation (6.32) for the distribution ρ and a bounded function h has as its solution*

$$\begin{aligned} f_h(x) &= \frac{1}{s(x+1)p_{x+1}} \sum_{i=a}^{x} (h(i) - E[h(Z)])p_i \qquad (6.33)\\ &= \frac{-1}{s(x+1)p_{x+1}} \sum_{i=x+1}^{b} (h(i) - E[h(Z)])p_i, \end{aligned}$$

when $a \le x < b$ and $f_h = 0$ elsewhere. Furthermore, this f_h belongs to C_3.

Proof: The proof is completely of the same structure as in the continuous case. First we check the proposed solution for the boundary points $x = a$ and $x = b$.

For $x = a$,

$$f_h(a) = \frac{(h(a) - E[h(Z)])p_a}{s(a+1)p_{a+1}},$$

and thus the left-hand side of (6.32) is equal to

$$\begin{aligned} & s(a)\nabla f_h(a) + \tau(a)f_h(a) \\ &= s(a)f_h(a) + \tau(a)f_h(a) \\ &= \frac{(h(a) - E[h(Z)])s(a)p_a}{s(a+1)p_{a+1}} + \frac{(h(a) - E[h(Z)])\tau(a)p_a}{s(a+1)p_{a+1}} \\ &= 0 + \frac{(h(a) - E[h(Z)])\Delta(s(a)p_a)}{s(a+1)p_{a+1}} \\ &= h(a) - E[h(Z)]. \end{aligned}$$

For $x = b$, we have

$$\begin{aligned} s(b)\nabla f_h(b) + \tau(b)f_h(b) &= -s(b)f(b-1) \\ &= (-s(b))\left(\frac{-(h(b) - E[h(Z)])p_b}{s(b)p_b}\right) \\ &= h(b) - E[h(Z)]. \end{aligned}$$

Next, note that for $a < x < b$,

$\nabla f_h(x)$

$$
\begin{aligned}
&= \frac{-\Delta(s(x)p_x)}{s(x)s(x+1)p_x p_{x+1}} \sum_{i=a}^{x}(h(i)-E[h(Z)])p_i + \frac{h(x)-E[h(Z)]}{s(x)} \\
&= \frac{-\tau(x)p_x}{s(x)s(x+1)p_x p_{x+1}} \sum_{i=a}^{x}(h(i)-E[h(Z)])p_i + \frac{h(x)-E[h(Z)]}{s(x)} \\
&= \frac{-\tau(x)}{s(x)s(x+1)p_{x+1}} \sum_{i=a}^{x}(h(i)-E[h(Z)])p_i + \frac{h(x)-E[h(Z)]}{s(x)}.
\end{aligned}
$$

Substituting the proposed solution (6.33) into the left-hand side of the Stein equation gives

$$
\begin{aligned}
& s(x)\nabla f_h(x) + \tau(x) f_h(x) \\
&= \frac{-\tau(x)}{s(x+1)p_{x+1}} \sum_{i=a}^{x}(h(i)-E[h(Z)])p_i + (h(x)-E[h(Z)]) \\
&\quad + \frac{\tau(x)}{s(x+1)p_{x+1}} \sum_{i=a}^{x}(h(i)-E[h(Z)])p_i \\
&= h(x) - E[h(Z)].
\end{aligned}
$$

The second expression for f_h follows from the fact

$$
\sum_{i=a}^{x}(h(i)-E[h(Z)])p_i + \sum_{i=x+1}^{b}(h(i)-E[h(Z)])p_i = 0.
$$

The proof that $f_h \in C_3$ is completely similar to that of the continuous case. This proves our proposition. ◇

The next step is to estimate

$$
E[s(W)\nabla f_h(W) + \tau(W) f_h(W)] \tag{6.34}
$$

and hence $E[h(W)] - E[h(Z)]$. To show the quantity in (6.34) is small, it is necessary to use the structure of W. In addition we might require certain smoothness conditions on f_h that would translate into smoothness conditions on h by the following lemma.

Let $\tau(x) = cx + d$, $c < 0$, and $l = -d/c$ be the only zero of τ. We need the following positive constant

$$
M = \frac{1}{p_{\lfloor l \rfloor+1} s(\lfloor l \rfloor + 1)} \max\left(\sum_{i=a}^{\lfloor l \rfloor} p_i, \sum_{i=\lfloor l \rfloor+1}^{b} p_i\right).
$$

Lemma 4 *Suppose h is a bounded function and let f_h be the solution of the Stein equation given by (6.32). Then*

$$
||f_h(x)|| \le M ||h(x) - E[h(Z)]||, \tag{6.35}
$$

where $||f(x)|| = \sup_{a \le x \le b} |f(x)|$.

Proof: Note that $\tau(x)$ is positive and decreasing in $[a, l)$ and negative and decreasing in $(l, b]$.

So we have for $a \le x < \lfloor l \rfloor$,

$$\begin{aligned} F(x) = \sum_{i=a}^{x} p_i &\le \sum_{i=a}^{x} \frac{\tau(i)}{\tau(x)} p_i \\ &= \frac{1}{\tau(x)} \sum_{i=a}^{x} \Delta(s(i)p_i) \\ &= \frac{s(x+1)p_{x+1}}{\tau(x)}. \end{aligned} \tag{6.36}$$

Similarly, for $\lfloor l \rfloor + 1 \le x < b$, we have

$$1 - F(x) = \sum_{i=x+1}^{b} p_i \le \frac{-s(x+1)p_{x+1}}{\tau(x)}. \tag{6.37}$$

Now, for $a \le x \le \lfloor l \rfloor$,

$$\begin{aligned} |f_h(x)| &= \left| \frac{1}{s(x+1)p_{x+1}} \sum_{i=a}^{x} (h(i) - E[h(Z)])p_i \right| \\ &\le \|h(x) - E[h(Z)]\| \frac{\sum_{i=a}^{x} p_i}{s(x+1)p_{x+1}}. \end{aligned} \tag{6.38}$$

Similarly for $\lfloor l \rfloor \le x < b$,

$$\begin{aligned} |f_h(x)| &= \left| \frac{-1}{s(x+1)p_{x+1}} \sum_{i=x+1}^{b} (h(i) - E[h(Z)])p_i \right| \\ &\le \|h(x) - E[h(Z)]\| \frac{\sum_{i=x+1}^{b} p_i}{s(x+1)p_{x+1}}. \end{aligned} \tag{6.39}$$

Proceeding as in the continuous case, we prove that the expressions

$$\frac{\sum_{i=a}^{x} p_i}{s(x+1)p_{x+1}} \quad \text{and} \quad \frac{\sum_{i=x+1}^{b} p_i}{s(x+1)p_{x+1}},$$

of (6.38) and (6.39) attain their maximum at $x = \lfloor l \rfloor$. To show this we calculate

$$\nabla \left(\frac{\sum_{i=a}^{x} p_i}{s(x+1)p_{x+1}} \right) = \frac{-\tau(x)}{p_{x+1}s(x)s(x+1)} \sum_{i=a}^{x} p_i + \frac{1}{s(x)}$$

and

$$\nabla \left(\frac{\sum_{i=x+1}^{b} p_i}{s(x+1)p_{x+1}} \right) = \frac{-\tau(x)}{p_{x+1}s(x)s(x+1)} \sum_{i=x+1}^{b} p_i - \frac{1}{s(x)}.$$

Next we use (6.36) and (6.37) to obtain for $a \le x \le \lfloor l \rfloor$,

$$\nabla\left(\frac{\sum_{i=a}^{x} p_i}{s(x+1)p_{x+1}}\right) \ge 0$$

and for $\lfloor l \rfloor + 1 \le x < b$,

$$\nabla\left(\frac{\sum_{i=x+1}^{b} p_i}{s(x+1)p_{x+1}}\right) \le 0.$$

In conclusion we have $||f_h(x)|| \le M||h - E[h(Z)]||$. ◇

Lemma 5 *Suppose h is a bounded function and let f_h be the solution of the Stein equation given by (6.32). Then*

$$||\nabla f_h(x)|| \le \max\left(\frac{1}{|\tau(a)|}, \sup_{a<x\le b}(2/s(x))\right)||h(x) - E[h(Z)]||. \tag{6.40}$$

Proof: For $x = a$, we have

$$\begin{aligned} |\nabla f_h(a)| &= \left|\frac{(h(a) - E[h(Z)])p_a}{s(a+1)p_{a+1}}\right| = \left|\frac{h(a) - E[h(Z)]}{\tau(a)}\right| \\ &\le \frac{||h(x) - E[h(Z)]||}{|\tau(a)|}. \end{aligned}$$

For $x = b$, we have

$$\begin{aligned} |\nabla f_h(b)| &= |f_h(b-1)| = \left|\frac{h(b) - E[h(Z)]}{s(b)}\right| \\ &\le \sup_{a<x\le b}(2/s(x))||h(x) - E[h(Z)]||. \end{aligned}$$

Because for $a < x < b$,

$$\begin{aligned} &\nabla f_h(x) \\ &= \frac{-\tau(x)}{s(x)s(x+1)p_{x+1}} \sum_{i=a}^{x}(h(i) - E[h(Z)])p_i + \frac{h(x) - E[h(Z)]}{s(x)} \\ &= \frac{\tau(x)}{s(x)s(x+1)p_{x+1}} \sum_{i=x+1}^{b}(h(i) - E[h(Z)])p_i + \frac{h(x) - E[h(Z)]}{s(x)}, \end{aligned}$$

we have for $a < x \le \lfloor l \rfloor$,

$$\begin{aligned} |\nabla f_h(x)| &\le ||h(x) - E[h(Z)]||\left(\frac{\tau(x)F(x)}{s(x+1)s(x)P_{x+1}} + \frac{1}{s(x)}\right) \\ &\le ||h(x) - E[h(Z)]||\left(\frac{2}{s(x)}\right), \end{aligned}$$

where we used (6.36) for the last inequality.

Similarly, for $\lfloor l \rfloor + 1 \leq x < b$ we have

$$\begin{aligned} |\nabla f_h(x)| &\leq \|h(x) - E[h(Z)]\| \left(\frac{-\tau(x)(1-F(x))}{s(x)s(x+1)p_{x+1}} + \frac{1}{s(x)} \right) \\ &\leq \|h(x) - E[h(Z)]\| \left(\frac{2}{s(x)} \right). \end{aligned}$$

This ends the proof. ◇

Examples

1. The Poisson distribution $P(\mu)$ is given by the probabilities

$$p_x = e^{-\mu} \frac{\mu^x}{x!}, \quad x \in \{0, 1, 2, \ldots\}.$$

An easy calculation gives $s(x) = x$ and $\tau(x) = \mu - x$. So the Stein operator for the Poisson distribution $P(\mu)$ is given by

$$\begin{aligned} Af(x) &= x\nabla f(x) + (\mu - x)f(x) \\ &= \mu f(x) - xf(x-1), \end{aligned}$$

which is the same as in (6.8). This case was studied by [23] and many others [9], [10], [11], [12].

2. The binomial distribution $\text{Bin}(N, p)$ on $\{0, 1, 2, \ldots, N\}$ with parameter $0 < p < 1$ is given by the probabilities

$$p_x = \binom{N}{x} p^x q^{N-x}, \quad x \in \{0, 1, 2, \ldots, N\},$$

where $q = 1 - p$. An easy calculation gives $s(x) = qx$ and $\tau(x) = pN - x$. So the Stein operator for the $\text{Bin}(N, p)$ distribution is given by

$$\begin{aligned} Af(x) &= qx\nabla f(x) + (pN - x)f(x) \\ &= p(N - x)f(x) - qxf(x-1). \end{aligned}$$

3. The Pascal distribution $\text{Pa}(\alpha, p)$ with parameters $\alpha > 0$ and $0 < p < 1$ is given by

$$p_x = \binom{x + \alpha - 1}{x} p^\alpha q^x, \quad x \in \{0, 1, 2, \ldots\},$$

where $q = 1 - p$. An easy calculation gives $s(x) = x$ and $\tau(x) = q\alpha - px$. So the Stein operator for the $\text{Pa}(\alpha, p)$ distribution is given by

$$\begin{aligned} Af(x) &= x\nabla f(x) + (q\alpha - px)f(x) \\ &= q(\alpha + x)f(x) - xf(x-1). \end{aligned}$$

4. The hypergeometric distribution HypII(α, β, N), with parameters $\alpha \geq N$, $\beta > N$, and N a nonnegative integer, is given by

$$p_x = \frac{\binom{\alpha}{x}\binom{\beta}{N-x}}{\binom{\alpha+\beta}{N}}, \qquad x \in \{0, 1, 2, \ldots, N\}.$$

An easy calculation gives $s(x) = x(\beta - N + x)$ and $\tau(x) = \alpha N - (\alpha + \beta)x$. So the Stein operator for the HypII(α, β, N) distribution is given by

$$\begin{aligned} \mathcal{A}f(x) &= x(\beta - N + x)\nabla f(x) + (\alpha N - (\alpha + \beta)x)f(x) \\ &= (N - x)(\alpha - x)f(x) - x(\beta - N + x)f(x - 1). \end{aligned}$$

This Stein equation seems to be news.

6.4.3 Orthogonal Polynomials

After having considered the close relation between the defining difference and differential equations of the involved distributions and their Stein(-Markov) operators, it is a natural question to look for the role of the related orthogonal polynomials. The key link in the continuous case is the differential equation of hypergeometric type (1.6) which is satisfied by the classical orthogonal polynomials of a continuous variable

$$s(x)y'' + \tau(x)y' + \lambda y = 0,$$

where $s(x)$ and $\tau(x)$ are polynomials of at most second and first degree, respectively, and λ is a constant. In the discrete case, the link is the difference equation of hypergeometric type (1.9) which is satisfied by the classical orthogonal polynomials of a discrete variable

$$s(x)\Delta\nabla y(x) + \tau(x)\Delta y(x) + \lambda y(x) = 0,$$

where $s(x)$ and $\tau(x)$ are again polynomials of at most second and first degree, respectively, and λ is a constant.

Let $Q_n(x)$ be the orthogonal polynomials with respect to the distribution ρ; then the $Q_n(x)$ satisfy equations of hypergeometric type for some specific constants $\lambda_n \neq 0$. But this means that we have

$$\mathcal{A}Q_n(x) = -\lambda_n Q_n(x). \tag{6.41}$$

In this way we can formally solve the Stein–Markov equation

$$\mathcal{A}f = h(x) - E[h(Z)] \tag{6.42}$$

with the aid of orthogonal polynomials. Let $F(x) = \Pr(Z \leq x)$, the involved distribution function, and the $d_n \neq 0$ as in (1.1). Suppose

$$h(x) - E[h(Z)] = \sum_{n=0}^{\infty} a_n Q_n(x),$$

where we can determine the a_n by

$$a_n = \int_S Q_n(x)(h(x) - E[h(Z)])dF(x)/d_n^2, \quad n \geq 0,$$

with S the support of the distribution ρ.

Note that

$$a_0 = Q_0(x) \int_S (h(x) - E[h(Z)])dF(x) = 0.$$

But then for a given h the solution of (6.42) is given by

$$f_h(x) = \sum_{n=1}^{\infty} \frac{-a_n}{\lambda_n} Q_n(x).$$

Indeed, we have

$$\begin{aligned}
\mathcal{A} f_h(x) &= \mathcal{A} \sum_{n=1}^{\infty} \frac{-a_n}{\lambda_n} Q_n(x) \\
&= \sum_{n=1}^{\infty} \frac{-a_n}{\lambda_n} \mathcal{A} Q_n(x) \\
&= \sum_{n=1}^{\infty} a_n Q_n(x) \\
&= h(x) - E[h(Z)].
\end{aligned}$$

6.4.4 Markov Process

Another place where the orthogonal polynomials appear is in Barbour's operator method. Recall that we are considering some distribution ρ, continuous or discrete, together with a Stein–Markov operator $\mathcal{A}$ of a Markov process, X_t say.

Discrete Case

In the discrete case the operator $\mathcal{A}$ has the form

$$\begin{aligned}
&\mathcal{A} f(x) \\
&= s(x)\Delta\nabla f(x) + \tau(x)\Delta f(x) \\
&= (s(x) + \tau(x))f(x+1) - (2s(x) + \tau(x))f(x) + s(x)f(x-1),
\end{aligned}$$

which is the operator of a birth and death process with birth and death rates

$$\kappa_n = s(n) + \tau(n) \qquad \text{and} \qquad \mu_n = s(n),$$

respectively, if $\kappa_n, \mu_n \geq 0$.

The orthogonal polynomials $Q_n(x)$ of ρ satisfy

$$\begin{aligned} &\mathcal{A}Q_n(x) \\ &\quad= (s(x)+\tau(x))Q_n(x+1) - (2s(x)+\tau(x))Q_n(x) + s(x)Q_n(x-1) \\ &\quad= -\lambda_n Q_n(x). \end{aligned} \tag{6.43}$$

Suppose we have a duality relation of the form

$$Q_n(x) = \hat{Q}_x(\lambda_n)$$

and that $\hat{Q}_x$ is a polynomial of degree x.

Then (6.43) can be written as

$$\begin{aligned} &-\lambda_n \hat{Q}_x(\lambda_n) = \\ &\quad (s(x)+\tau(x))\hat{Q}_{x+1}(\lambda_n) - (2s(x)+\tau(x))\hat{Q}_x(\lambda_n) + s(x)\hat{Q}_{x-1}(\lambda_n). \end{aligned}$$

Interchanging the role of x and n we clearly see that this results in a three-term recurrence equation

$$\begin{aligned} &-\lambda_x \hat{Q}_n(\lambda_x) = \\ &\quad (s(n)+\tau(n))\hat{Q}_{n+1}(\lambda_x) - (2s(n)+\tau(n))\hat{Q}_n(\lambda_x) + s(n)\hat{Q}_{n-1}(\lambda_x). \end{aligned}$$

By Favard's Theorem the $\hat{Q}_n$ must be orthogonal polynomials with respect to some distribution, $\tilde{\rho}$ say. Furthermore, note that these polynomials are the birth–death polynomials of the birth and death process X_t. According to the Karlin and McGregor spectral representation (3.2) we have

$$P_{ij}(t) \equiv \Pr(X_t = j | X_0 = i) = \pi_j \int_0^\infty e^{-\lambda_y t} \hat{Q}_i(\lambda_y)\hat{Q}_j(\lambda_y) d\tilde{F}(y), \tag{6.44}$$

where

$$\pi_0 = 1 \quad \text{and} \quad \pi_j = \frac{\kappa_0 \kappa_1 \dots \kappa_{j-1}}{\mu_1 \mu_2 \dots \mu_j}, \qquad j \geq 1,$$

and $\tilde{F}(x)$ is the distribution function of $\tilde{\rho}$.

The stationary distribution is given by

$$r_i = \frac{\pi_i}{\sum_{k=0}^\infty \pi_k}.$$

Note that this distribution is completely defined by the fraction of successive probabilities

$$\frac{r_{i+1}}{r_i} = \frac{\kappa_i}{\mu_{i+1}} = \frac{\sigma(i)+\tau(i)}{\sigma(i+1)}.$$

Comparing this with (1.12) we see that the stationary distribution is indeed our starting distribution ρ.

We now work out this procedure for some well-known discrete distributions.

Examples

1. The Poisson distribution $\mathrm{P}(\mu)$ has a Stein–Markov operator $\mathcal{A}$, given by
$$\mathcal{A}f(x) = \mu f(x+1) - (x+\mu)f(x) + xf(x-1).$$
This is the operator of a birth and death process on $\{0, 1, 2 \dots\}$ with birth and death rates
$$\kappa_n = \mu \quad \text{and} \quad \mu_n = n, \qquad n \geq 0,$$
respectively. This birth and death process can be seen as an immigration–death process with a constant immigration rate μ and unit per capita death rate. For more details see also the example with the $M/M/\infty$ queue in Chapter 3. The birth–death polynomials $Q_n(x)$ for this process are recursively defined by the relations
$$-xQ_n(x) = \mu Q_{n+1}(x) - (\mu + n)Q_n(x) + nQ_{n-1}(x), \tag{6.45}$$
together with $Q_0(x) = 1$ and $Q_{-1}(x) = 0$.
The polynomials that are orthogonal with respect to the Poisson distribution $\mathrm{P}(\mu)$ are the Charlier polynomials $C_n(x; \mu)$. The Charlier polynomials satisfy the following hypergeometric type equation
$$-nC_n(x; \mu) = \mu C_n(x+1; \mu) - (\mu + n)C_n(x; \mu) + nC_n(x-1; \mu)$$
and are self-dual; i.e.,
$$C_n(x; \mu) = C_x(n; \mu).$$
Using this duality relation we obtain the three-term recurrence relation of the Charlier polynomials
$$-nC_x(n; \mu) = \mu C_{x+1}(n; \mu) - (\mu + n)C_x(n; \mu) + nC_{x-1}(n; \mu).$$
But this is after interchanging the roles of x and n exactly of the same form as (6.45); so we conclude that
$$Q_n(x) = C_n(x; \mu).$$
In this way, using Karlin and McGregor's spectral representation (3.2), we can express the transition probabilities of our process X_t as
$$P_{ij}(t) = \Pr(X_t = j | X_0 = i) = \pi_j \sum_{x=0}^{\infty} e^{-xt} C_i(x; \mu) C_j(x; \mu) e^{-\mu} \frac{\mu^x}{x!},$$
where $\pi_j = \mu^j / j!$.

2. The Pascal distribution $\mathrm{Pa}(\gamma, \mu)$ has a Stein–Markov operator $\mathcal{A}$, given by
$$\mathcal{A}f(x) = \mu(x+\gamma)f(x+1) - (\mu(x+\gamma)+x)f(x) + xf(x-1).$$
This is the operator of a birth and death process on $\{0, 1, 2 \ldots\}$ with birth and death rates
$$\kappa_n = \mu(n+\gamma) \quad \text{and} \quad \mu_n = n, \qquad n \geq 0,$$
respectively. This birth and death process can be seen as a linear birth and death process with a linear birth rate $\mu(x+\gamma)$ and unit per capita death rate. For more details see also the example with the linear birth and death process in Chapter 3. The birth–death polynomials $Q_n(x)$ for this process are recursively defined by the relations
$$-xQ_n(x) = \mu(n+\gamma)Q_{n+1}(x) - (\mu(n+\gamma)+n)Q_n(x) + nQ_{n-1}(x), \tag{6.46}$$
together with $Q_0(x) = 1$ and $Q_{-1}(x) = 0$.
The polynomials that are orthogonal with respect to the Pascal distribution $\mathrm{Pa}(\gamma, \mu)$ are the Meixner polynomials $M_n(x; \gamma, \mu)$. The Meixner polynomials satisfy the following hypergeometric type equation
$$\begin{aligned} &-nM_n(x;\gamma,\mu) \\ &= \mu(n+\gamma)M_n(x+1;\gamma,\mu) - (\mu(n+\gamma)+n)M_n(x;\gamma,\mu) \\ &\qquad +nM_n(x-1;\gamma,\mu) \end{aligned}$$
and are self-dual; i.e.,
$$M_n(x;\gamma,\mu) = M_x(n;\gamma,\mu).$$
Using this duality relation we obtain the three-term recurrence relation of the Meixner polynomials
$$\begin{aligned} &-nM_x(n;\gamma,\mu) = \\ &\quad \mu(n+\gamma)M_{x+1}(n;\gamma,\mu) - (\mu(n+\gamma)+n)M_x(n;\gamma,\mu) \\ &\qquad +nM_{x-1}(n;\gamma,\mu). \end{aligned}$$
But this is after interchanging the roles of x and n exactly of the same form as (6.46), so we conclude that
$$Q_n(x) = M_n(x;\gamma,\mu).$$
In this way, using Karlin and McGregor's spectral representation (3.2), we can express the transition probabilities of our process X_t as
$$\begin{aligned} &P_{ij}(t) = \Pr(X_t = j | X_0 = i) = \\ &\quad \pi_j \sum_{x=0}^{\infty} e^{-xt} M_i(x;\gamma,\mu) M_j(x;\gamma,\mu)(1-\mu)^\gamma \mu^x (\gamma)_x / x!, \end{aligned}$$

where $\pi_j = \mu^j(\gamma)_j/j!$.

3. The binomial distribution $\text{Bin}(N,p)$ has a Stein–Markov operator $\mathcal{A}$, given by

$$\mathcal{A}f(x) = p(N-x)f(x+1) - (p(N-x)+qx)f(x) + qxf(x-1),$$

where $q = 1-p$. This is the operator of a birth and death process on $\{0,1,2\ldots,N\}$ with birth and death rates

$$\kappa_n = p(N-n) \quad \text{and} \quad \mu_n = qn, \qquad 0 \le n \le N,$$

respectively. The birth–death polynomials $Q_n(x)$ for this process are recursively defined by the relations

$$-xQ_n(x) = p(N-n)Q_{n+1}(x) - (p(N-n)+qn)Q_n(x) + qnQ_{n-1}(x), \tag{6.47}$$

together with $Q_0(x) = 1$ and $Q_{-1}(x) = 0$.

The polynomials that are orthogonal with respect to the binomial distribution $\text{Bin}(N,p)$ are the Krawtchouk polynomials $K_n(x;N,p)$. The Krawtchouk polynomials satisfy the following hypergeometric type equation

$$\begin{aligned}-nK_n(x;N,p) &= \\ &p(N-n)K_n(x+1;N,p) - (p(N-n)+qn)K_n(x;N,p) \\ &\quad + qnK_n(x-1;N,p)\end{aligned}$$

and are self-dual; i.e.,

$$K_n(x;N,p) = K_x(n;N,p).$$

Using this duality relation we obtain the three-term recurrence relation of the Krawtchouk polynomials

$$\begin{aligned}-nK_x(n;\gamma,\mu) &= \\ &p(N-n)K_{x+1}(n;N,p) - (p(N-n)+qn)K_x(n;N,p) \\ &\quad + qnK_{x-1}(n;N,p).\end{aligned}$$

But this is after interchanging the roles of x and n exactly of the same form as (6.47), so we conclude that

$$Q_n(x) = K_n(x;N,p).$$

In this way, using Karlin and McGregor's spectral representation (3.2), we can express the transition probabilities of our process X_t as

$$\begin{aligned}P_{ij}(t) = \Pr(X_t = j|X_0 = i) &= \\ &\pi_j \sum_{x=0}^{N} e^{-xt} K_i(x;N,p)K_j(x;N,p)\binom{N}{x}p^x q^{N-x},\end{aligned}$$

where $\pi_j = \binom{N}{j} p^j q^{-j}$.

4. The hypergeometric distribution HypII(α, β, N) has a Stein–Markov operator $\mathcal{A}$, given by

$$\begin{aligned}\mathcal{A}f(x) & \\ = & \ (N-x)(\alpha-x)f(x+1) \\ & -((N-x)(\alpha-x)+x(\beta-N+x))f(x) \\ & +x(\beta-N+x)f(x-1).\end{aligned}$$

This is the operator of a birth and death process on $\{0,1,2\ldots,N\}$ with quadratic birth and death rates

$$\kappa_n = (N-n)(\alpha-n) \quad \text{and} \quad \mu_n = n(\beta-N+n), \qquad 0 \le n \le N,$$

respectively. The birth–death polynomials $Q_n(x)$ for this process are recursively defined by the relations

$$\begin{aligned}-xQ_n(x) &= (N-n)(\alpha-n)Q_{n+1}(x) \\ &\quad -((N-n)(\alpha-n)+n(\beta-N+n))Q_n(x) \\ &\quad +n(\beta-N+n)Q_{n-1}(x), \end{aligned} \qquad (6.48)$$

together with $Q_0(x) = 1$ and $Q_{-1}(x) = 0$.

The polynomials that are orthogonal with respect to the hypergeometric distribution HypII(α, β, N) are the famous Hahn polynomials $Q_n(x; -\alpha-1, -\beta-1, N)$. The Hahn polynomials $Q_n(x; -\alpha-1, -\beta-1, N)$ satisfy the following hypergeometric type equation

$$\begin{aligned}& n(n-\alpha-\beta-1)Q_n(x; -\alpha-1, -\beta-1, N) = \\ & \quad (N-x)(\alpha-x)Q_n(x+1; -\alpha-1, \beta-1, N) \\ & \quad -((N-x)(\alpha-x)+x(x+\beta-N))Q_n(x; -\alpha-1, -\beta-1, N) \\ & \quad +x(x+\beta-N)Q_n(x-1; -\alpha-1, -\beta-1, N).\end{aligned}$$

Furthermore, we have from the duality relation (1.17),

$$Q_n(x; -\alpha-1, -\beta-1, N) = R_x(\lambda_n; -\alpha-1, -\beta-1, N),$$

where the R_x are the dual Hahn polynomials and $\lambda_n = n(-\alpha-\beta-1)$. In what follows for notational convenience we often write $R_x(\lambda_n)$ instead of $R_x(\lambda_n; -\alpha-1, -\beta-1, N)$.

Using this duality relation we obtain the three-term recurrence relation of the dual Hahn polynomials

$$\begin{aligned}& \lambda_n R_x(\lambda_n) \\ & \quad = (N-x)(\alpha-x)R_{x+1}(\lambda_n) \\ & \qquad -((N-x)(\alpha-x)+x(x+\beta-N))R_x(\lambda_n) \\ & \qquad +x(x+\beta-N)R_{x-1}(\lambda_n).\end{aligned}$$

But this is after interchanging the roles of x and n of the same form as (6.48); so we conclude that

$$Q_n(x) = R_n(-x; -\alpha - 1, -\beta - 1, N).$$

In this way, using Karlin and McGregor's spectral representation (3.2), we can express the transition probabilities of our process X_t as

$$P_{ij}(t) = \Pr(X_t = j | X_0 = i) = \pi_j \sum_{x=0}^{N} e^{\lambda_x t} R_i(\lambda_x) R_j(\lambda_x) \tilde{\rho}(x),$$

where

$$\tilde{\rho}(x) = \frac{(N!)(-N)_x(-\alpha)_x(2x-\alpha-\beta-1)\binom{N-\alpha-1}{N}}{(-1)^x(x!)(-\beta)_x(x-\alpha-\beta-1)_{N+1}}$$

and

$$\pi_j = \frac{\binom{\alpha}{j}\binom{\beta}{N-j}}{\binom{\beta}{N}}.$$

Continuous Case

In the continuous case the operator $\mathcal{A}$ has the form

$$\mathcal{A}f(x) = s(x)f''(x) + \tau(x)f'(x)$$

which is the operator of a diffusion with drift coefficient $\mu(x) = \tau(x)$ and diffusion coefficient $\sigma^2(x) = 2s(x) > 0$.

Recall that the polynomials $y_n(x)$, which are orthogonal with respect to ρ, satisfy

$$\mathcal{A}y_n(x) = -\lambda_n y_n(x), \quad n \geq 0.$$

In Chapter 2, Section 6 we look at some general diffusions. If in addition we suppose that $\tau(x)$ and $s(x)$ are polynomials of at most second degree and first degree, respectively, and we look for eigenfunctions as in (2.11) we clearly see that the polynomials, $y_n(x)$, which are orthogonal with respect to ρ, are eigenfunctions corresponding to the eigenvalue $-\lambda_n$ of the operator $\mathcal{A}$. Furthermore the eigenfunctions are complete and we have a spectral representation as in Chapter 2, Section 6.2.

Examples

1. The standard normal distribution N(0, 1) and the Ornstein–Uhlenbeck process. Suppose we have $\mu(x) = \tau(x) = -x$ and $\sigma^2(x) = 2s(x) = 2$; then we have

$$\mathcal{A}H_n(x/\sqrt{2}) = -nH_n(x/\sqrt{2}),$$

where the operator A is given by

$$\mathcal{A}f = f''(x) - xf'(x).$$

In Chapter 2 we found the spectral representation, in terms of Hermite polynomials, for the transition density

$$p(t;x,y) = \frac{e^{-y^2/2}}{\sqrt{2\pi}} \sum_{n=0}^{\infty} e^{-nt} H_n(x/\sqrt{2}) H_n(y/\sqrt{2}) \frac{1}{2^n n!}. \tag{6.49}$$

2. The Gamma distribution $G(r, \lambda^{-1})$ and the Laguerre diffusion. Suppose we have $\mu(x) = \tau(x) = r - \lambda x$ and $\sigma^2(x) = 2s(x) = 2x$, where $0 < x < \infty$ and the constants satisfy $\lambda, r > 0$; then

$$p(t;x,y) = \frac{\lambda^r y^{r-1} e^{-\lambda y}}{\Gamma(r)} \sum_{n=0}^{\infty} e^{-n\lambda t} L_n^{(r-1)}(\lambda x) L_n^{(r-1)}(\lambda y) \frac{\Gamma(n+1)}{\Gamma(n+r)},$$

with as always L_n the Laguerre polynomial.

3. The Beta distribution $B(\alpha, \beta)$ and the Jacobi diffusion. Suppose we have $\mu(x) = \tau(x) = (\alpha - (\alpha+\beta)x)/2$ and $\sigma^2(x) = (1-x)x$. In this case the spectral expansion is in terms of the Jacobi polynomials

$$p(t;x,y) = \frac{y^{\alpha-1}(1-y)^{\beta-1}}{B(\alpha,\beta)} \times$$

$$\sum_{n=0}^{\infty} e^{-n(n+\alpha+\beta-1)t/2} P_n^{(\beta-1,\alpha-1)}(2x-1) P_n^{(\beta-1,\alpha-1)}(2y-1) \pi_n,$$

where

$$\pi_n = \frac{B(\alpha,\beta)(2n+\alpha+\beta-1)n!\Gamma(n+\alpha+\beta-1)}{\Gamma(n+\alpha)\Gamma(n+\beta)}.$$

Notes

Barbour's [8] generator method plays a key role in Stein's theory and uses Markov processes in the analysis of the approximation. In this chapter, the use of orthogonal polynomials in the Stein and Barbour methods was introduced. After the theoretical aspects, covered in this work, one could start looking at concrete models and try to obtain good approximations by using the properties of the orthogonal polynomials in the spectral representation of the Markov processes.

In [13] the Stein equation is also studied for the compound Poisson case. For a brief history of the developments and some possible prospects of Stein's method, we refer to [24]. In [4] a variety of examples of the wide applicability and utility of method in the case of Poisson approximation is given.

Conclusion

In this book we tried to give a probabilistic interpretation of the major part of the Askey scheme. Orthogonal polynomials of this scheme are related to stochastic processes: birth and death processes, random walks, Lévy processes, and diffusions. The relationships between the polynomials and the processes are of a wide variety. We discussed spectral representation relations for birth and death processes, random walks, and diffusions in different contexts. They appeared in the study of the time-dependent and asymptotic behavior of the processes and in Stein's method for the approximation of distributions. Martingale and stochastic integral relations for Lévy processes and sums of i.i.d. random variables formed another important part of this work.

Birth and death processes and their discrete counterparts, random walks, were studied with the aid of a spectral representation in terms of orthogonal polynomials. Due to the difficulties involved in analytical methods, it is almost impossible to find closed-form solutions of the transition functions of birth and death processes with complicated birth and death rates. The Karlin and McGregor representation [59] [74] of the transition probabilities, which uses a system of orthogonal polynomials, now called birth–death polynomials, satisfying a three-term recurrence relation involving the birth and death rates, is thus very useful for understanding the behavior of the birth and death process. We showed how these polynomials appear in some important distributions, namely, the (doubly) limiting conditional distributions. The use of the Karlin and McGregor representation in the analysis of the doubly limiting conditional distribution seems to be new. The representation has been already used for the limiting conditional distribution in [68]. Although in this work we only encountered birth–death polynomials that are part of the Askey scheme, the results are not restricted to this scheme and birth–death polynomials not in the Askey scheme can appear. In [88], for example, the so-called more-the-merrier birth and death process is analyzed and the (doubly) limiting conditional distributions are given in terms of orthogonal polynomials related to the Roger–Ramanujan continued fraction.

Furthermore birth and death processes and diffusions also appear in a completely different context: in Stein's approximation theory. Stein's method provides a way of finding approximations to the distribution of a random variable, which at the same time gives estimates of the approximation error involved. The strengths of the method are that it can be applied in many circumstances in which dependence plays a part. A key tool in Stein's theory is the generator method developed by Barbour [10], which makes use of certain Markov processes. For a given distribution there may be various Markov processes which fit in Barbour's method. However, up to now, it has not been clear which Markov process to take to obtain good results. We showed how for a broad class of distributions there is a special Markov process, a birth and death process or a diffusion, that takes a leading role in the analysis. The spectral representation of the transition probabilities of this Markov process will be in terms of orthogonal polynomials closely related to the distribution to be approximated. This systematic treatment together with the introduction of orthogonal polynomials in the analysis seems to be new. Furthermore some earlier uncovered examples like the Beta, the Student's t, and the hypergeometric distribution have now been worked out. Now that we found this relation we could start analyzing concrete situations and discover if we could obtain nice approximation bounds based on the properties of the orthogonal polynomials involved.

Another important part of this manuscript is the study of special martingale relations. We established a new connection between the class of Sheffer polynomials and Lévy processes (or sums of i.i.d. random variables). Lévy processes appear in many areas, such as in models for queues, insurance risks, and more recently in mathematical finance. Some already well-known martingales appeared as special cases (Hermite and Charlier), but most of the martingales obtained were completely new. Also a new Lévy process was born, the Meixner process. It appears in the study of the Meixner–Pollaczek polynomials and has many nice properties that resemble some properties of risky assets. Applications of the Meixner process in mathematical finance already have been started [52] and look very promising.

Some of the Lévy–Sheffer martingales and their associated stochastic process play an even more spectacular role in stochastic integration theory. The Hermite polynomials are the stochastic counterparts of the regular monomials (of the classical deterministic integration) for stochastic integration theory with respect to Brownian motion. The same is true for the

Charlier polynomials and the Poisson process. These stochastic integral relations were already known by many people [56] [87]. In this work we proved that the Krawtchouk polynomials also play such a role in the stochastic integration (summation) theory with respect to the binomial process. The binomial process is an important model for the binary market [41]. Note that from these integration results, the martingale property for these polynomials and processes immediately follows.

The only normal martingales that possess the chaotic representation property and the weaker predictable representation property and which are at the same time also Lévy processes, are in essence Brownian motion and the compensated Poisson process. For a general Lévy process (satisfying some moment conditions), a more general chaotic representation for every square integral random variable in terms of these orthogonalized Teugels martingales is given. A general predictable representation with respect to the same set of orthogonalized martingales of square integrable random variables and of square integrable martingales is an easy consequence of the chaotic representation.

Appendix A
Distributions

In the table of the discrete distributions, we always have that $0 < p < 1$ and $N \in \{0, 1, \ldots\}$, for the Poisson distribution we have $\mu > 0$, and for the Pacal distribution we have $\gamma > 0$ and $0 < \mu < 1$.

In the table of continuous distributions, we always have that $\sigma^2 > 0, \alpha, \beta, \lambda > 0$, $a < b$, and $n \in \{1, 2, \ldots\}$.

TABLE A.1. Discrete Distributions

Name	Notation	Probabilities	Support
Bernoulli	$\mathrm{B}(p)$	p if $x = 0$, $1 - p$ if $x = 1$	$\{0, 1\}$
Poisson	$\mathrm{P}(\mu)$	$e^{-\mu}\mu^x/x!$	$\{0, 1, 2, \ldots\}$
Pascal	$\mathrm{Pa}(\gamma, \mu)$	$\frac{(1-\mu)^\gamma(\gamma)_x\mu^x}{x!}$	$\{0, 1, 2, \ldots\}$
Binomial	$\mathrm{Bin}(N, p)$	$\binom{N}{x}p^x(1-p)^{N-x}$	$\{0, 1, \ldots, N\}$
Geometric	$\mathrm{Geo}(p)$	$p(1-p)^x$	$\{0, 1, 2, \ldots\}$
Hypergeo-metric I	$\mathrm{HypI}(\alpha, \beta, N)$	$\binom{N}{x}\frac{(\alpha+1)_x(\beta+1)_{N-x}}{(\alpha+\beta+2)_N}$	$\{0, 1, \ldots, N\}$
Hypergeo-metric II	$\mathrm{HypII}(\alpha, \beta, N)$	$\frac{\binom{\alpha}{x}\binom{\beta}{N-x}}{\binom{\alpha+\beta}{N}}$	$\{0, 1, \ldots, N\}$

TABLE A.2. Continuous Distributions

Name	Notation	Density	Support
Normal	$\mathrm{N}(\mu, \sigma^2)$	$\frac{1}{\sqrt{2\pi\sigma^2}}e^{-(x-\mu)^2/(2\sigma^2)}$	$\mathbb{R}$
Gamma	$\mathrm{G}(\alpha, \beta)$	$\frac{1}{\Gamma(\alpha)\beta^\alpha}x^{\alpha-1}e^{-x/\beta}$	$(0, \infty)$
Beta	$\mathrm{B}(\alpha, \beta)$	$\frac{1}{B(\alpha,\beta)}x^{\alpha-1}(1-x)^{\beta-1}$	$(0, 1)$
Exponen-tial	$\mathrm{Exp}(\lambda)$	$\lambda e^{-\lambda x}$	$(0, \infty)$
Uniform	$\mathrm{U}(a, b)$	$\frac{1}{b-a}$	(a, b)
Student's t	t_n	$\frac{\Gamma((n+1)/2)}{\sqrt{n\pi}\Gamma(n/2)}\left(1 + \frac{x^2}{n}\right)^{-(n+1)/2}$	$\mathbb{R}$

Appendix B

Tables of Classical Orthogonal Polynomials

Here we summarize the ingredients of the classical orthogonal polynomial $y_n(x)$ of degree n, which satisfies in the continuous case

$$s(x)y_n''(x) + \tau(x)y_n'(x) + \lambda_n y_n(x).$$

We have the following orthogonality relations

$$\int_S y_n(x)y_m(x)\rho(x)dx = d_n^2\delta_{nm},$$

where S is the support of $\rho(x)$.

In the discrete case, $y_n(x)$ satisfies

$$s(x)\Delta\nabla y_n(x) + \tau(x)\Delta y_n(x) + \lambda_n y_n(x).$$

We have the following orthogonality relations

$$\sum_{x\in S} y_n(x)y_m(x)\rho(x) = d_n^2\delta_{nm},$$

where S is the support of $\rho(x)$.

The constant a_n is the leading coefficient of $y_n(x)$ and the other constants b_n, γ_n, and c_n appear in the three-term recurrence relation

$$-xy_n(x) = b_n y_{n+1}(x) + \gamma_n y_n(x) + c_n y_{n-1}(x), \quad n \geq 0,$$

with initial conditions $y_{-1}(x) = 0$ and $y_0(x) = 1$.

B.1 Hermite Polynomials and the Normal Distribution

TABLE B.1. Hermite Polynomials and the Normal Distribution N(0, 1/2)

	Hermite — N(0, 1/2)
Notation	$H_n(x)$
Restrictions	
Hypergeometric Function	$(2x)^n \ {}_2F_0(-n/2, -(n-1)/2;;-1/x^2)$
Generating Function	$\sum_{n=0}^{\infty} H_n(x)\frac{z^n}{n!} = \exp(2xz - z^2)$
$\rho(x)$	$\exp(-x^2)/\sqrt{\pi}$
Support	$(-\infty, +\infty)$
$s(x)$	1
$\tau(x)$	$-2x$
λ_n	$2n$
a_n	2^n
b_n	$-1/2$
c_n	$-n$
γ_n	0
d_n^2	$2^n n!$

B.2 Scaled Hermite Polynomials and the Standard Normal Distribution

TABLE B.2. Scaled Hermite Polynomials and the Standard Normal Distribution N(0, 1)

	Scaled Hermite — N(0, 1)
Notation	$H_n(x/\sqrt{2})$
Restrictions	
Hypergeometric Function	$(\sqrt{2}x)^n \; {}_2F_0(-n/2, -(n-1)/2;\, ; -2/x^2)$
Generating Function	$\sum_{n=0}^{\infty} H_n(x/\sqrt{2}) \frac{z^n}{n!} = \exp(\sqrt{2}xz - z^2)$
$\rho(x)$	$\exp(-x^2/2)/\sqrt{2\pi}$
Support	$(-\infty, +\infty)$
$s(x)$	1
$\tau(x)$	$-x$
λ_n	n
a_n	$2^{n/2}$
b_n	$-1/\sqrt{2}$
c_n	$-\sqrt{2}n$
γ_n	0
d_n^2	$2^n n!$

B.3 Hermite Polynomials with Parameter and the Normal Distribution

TABLE B.3. Hermite Polynomials and the Standard Normal Distribution $N(0, t)$

	Hermite with Parameter — $\mathbf{N}(0, t)$
Notation	$H_n(x; t) = H_n(x/\sqrt{2t})$
Restrictions	$t > 0$
Hypergeometric Function	$(\sqrt{2/t}x)^n\ {}_2F_0(-n/2, -(n-1)/2;; -2t/x^2)$
Generating Function	$\sum_{n=0}^{\infty} H_n(x/\sqrt{2t}) \frac{z^n}{n!} = \exp(\sqrt{2}xz/\sqrt{t} - z^2)$
$\rho(x)$	$\exp(-x^2/(2t))/\sqrt{2\pi t}$
Support	$(-\infty, +\infty)$
$s(x)$	t
$\tau(x)$	$-x$
λ_n	n
a_n	$(\sqrt{2/t})^n$
b_n	$-\sqrt{t/2}$
c_n	$-\sqrt{2t}n$
γ_n	0
d_n^2	$2^n n!$

B.4 Charlier Polynomials and the Poisson Distribution

TABLE B.4. Charlier Polynomials and the Poisson Distribution P(μ)

	Charlier — P(μ)
Notation	$C_n(x;\mu)$
Restrictions	$\mu > 0$
Hypergeometric Function	${}_2F_0(-n,-x;;-1/\mu)$
Generating Function	$\sum_{n=0}^{\infty} C_n(x;\mu)\frac{z^n}{n!} = e^z(1-(z/\mu))^x$
$\rho(x)$	$e^{-\mu}\mu^x/x!$
Support	$\{0,1,2,\ldots\}$
$s(x)$	x
$\tau(x)$	$\mu - x$
λ_n	n
a_n	$(-1/\mu)^n$
b_n	μ
c_n	n
γ_n	$-(n+\mu)$
d_n^2	$\mu^{-n}n!$

B.5 Laguerre Polynomials and the Gamma Distribution

TABLE B.5. Laguerre Polynomials and the Gamma Distribution $\mathrm{G}(\alpha,1)$

	Laguerre — $\mathbf{G}(\alpha,1)$
Notation	$L_n^{(\alpha)}(x)$
Restrictions	$\alpha > -1$
Hypergeometric Function	$((\alpha+1)_n/n!)\ {}_1F_1(-n;\alpha+1;x)$
Generating Function	$\sum_{n=0}^{\infty} L_n^{(\alpha)}(x)z^n = (1-z)^{-\alpha-1}\exp(xz/(z-1))$
$\rho(x)$	$e^{-x}x^{\alpha}/\Gamma(\alpha+1)$
Support	$(0,+\infty)$
$s(x)$	x
$\tau(x)$	$\alpha-1-x$
λ_n	n
a_n	$(-1)^n/n!$
b_n	$n+1$
c_n	$n+\alpha$
γ_n	$-(2n+\alpha+1)$
d_n^2	$(\alpha+1)_n/n!$

B.6 Meixner Polynomials and the Pascal Distribution

TABLE B.6. Meixner Polynomials and the Pascal Distribution $\mathrm{Pa}(\gamma, \mu)$

	Meixner — Pa(γ, μ)
Notation	$M_n(x; \gamma, \mu)$
Restrictions	$0 < \mu < 1, \gamma > 0$
Hypergeometric Function	${}_2F_1(-n, -x; \gamma; 1 - (1/\mu))$
Generating Function	$\sum_{n=0}^{\infty} (\gamma)_n M_n(x; \gamma, \mu) \frac{z^n}{n!} = (1 - (z/\mu))^x (1 - z)^{-x-\gamma}$
$\rho(x)$	$(1 - \mu)^\gamma \mu^x (\gamma)_x / x!$
Support	$\{0, 1, 2, \ldots\}$
$s(x)$	$x/(1 - \mu)$
$\tau(x)$	$-x + \gamma\mu/(1 - \mu)$
λ_n	n
a_n	$(1 - (1/\mu))^n / (\gamma)_n$
b_n	$\mu(n + \gamma)/(1 - \mu)$
c_n	$n/(1 - \mu)$
γ_n	$-(n + \mu(n + \gamma))/(1 - \mu)$
d_n^2	$n! \mu^{-n} / (\gamma)_n$

B.7 Krawtchouk Polynomials and the Binomial Distribution

TABLE B.7. Krawtchouk Polynomials and the Binomial Distribution Bin(N,p)

	Krawtchouk — Bin(N,p)
Notation	$K_n(x; N, p)$
Restrictions	$0 < p < 1, p + q = 1, N \in \{0, 1, 2, \dots\}$
Hypergeometric Function	${}_2\tilde{F}_1(-n, -x; -N; 1/p)$
Generating Function	$\sum_{n=0}^{\infty} \binom{N}{n} K_n(x; N, p) z^n \simeq$ $(1 - (qz/p))^x (1 + z)^{N-x}$
$\rho(x)$	$\binom{N}{x} p^x q^{N-x}$
Support	$\{0, 1, 2, \dots, N\}$
$s(x)$	xq
$\tau(x)$	$pN - x$
λ_n	n
a_n	$p^{-n}/(-N)_n$
b_n	$p(N - n)$
c_n	nq
γ_n	$-(p(N - n) + nq)$
d_n^2	$n!(-1)^n (q/p)^n/(-N)_n$

B.8 Jacobi Polynomials and the Beta Kernel

TABLE B.8. Jacobi Polynomials and the Beta Kernel

	Jacobi — Beta
Notation	$P_n^{(\alpha,\beta)}(x)$
Restrictions	$\alpha, \beta > -1$
Hypergeometric Function	$\frac{(\alpha+1)_n}{n!}\ {}_2F_1\left(-n, n+\alpha+\beta+1; \alpha+1; \frac{1-x}{2}\right)$
Generating Function	
$\rho(x)$	$\frac{(1-x)^{\alpha}(1+x)^{\beta}}{2^{\alpha+\beta+1}B(\alpha+1,\beta+1)}$
Support	$(-1,1)$
$s(x)$	$(1-x^2)$
$\tau(x)$	$\beta-\alpha-x(\alpha+\beta+2)$
λ_n	$n(n+\alpha+\beta+1)$
a_n	$(n+\alpha+\beta+1)/(n!2^n)$
b_n	$-\frac{2(n+1)(n+\alpha+\beta+1)}{(2n+\alpha+\beta+1)(2n+\alpha+\beta+2)}$
c_n	$-\frac{2(n+\beta)(n+\alpha)}{(2n+\alpha+\beta)(2n+\alpha+\beta+1)}$
γ_n	$\frac{\alpha^2-\beta^2}{(2n+\alpha+\beta)(2n+\alpha+\beta+1)}$
d_n^2	$\frac{\Gamma(n+\beta+1)\Gamma(n+\alpha+1)}{B(\alpha+1,\beta+1)(2n+\alpha+\beta+1)n!\Gamma(n+\alpha+\beta+1)}$

B.9 Hahn Polynomials and the Hypergeometric Distribution

TABLE B.9. Hahn Polynomials and the Hypergeometric Distribution

	Hahn — HypI(α, β, N)
Notation	$Q_n(x; \alpha, \beta, N)$
Restrictions	$N \in \{0, 1, 2, \ldots\}$
Hypergeometric Function	${}_3\tilde{F}_2(-n, n+\alpha+\beta+1, -x; \alpha+1, -N; 1)$
Generating Function	
$\rho(x)$	$\binom{N}{x}\frac{(\alpha+1)_x(\beta+1)_{N-x}}{(\alpha+\beta+2)_N}$
Support	$\{0, 1, 2, \ldots, N\}$
$s(x)$	$x(N-x+\beta+1)$
$\tau(x)$	$N(\alpha+1) - x(\alpha+\beta+2)$
λ_n	$n(n+\alpha+\beta+1)$
a_n	$(n+\alpha+\beta+1)_n/((\alpha+1)_n(-N)_n)$
b_n	$\frac{(n+\alpha+\beta+1)(n+\alpha+1)(N-n)}{(2n+\alpha+\beta+1)(2n+\alpha+\beta+2)}$
c_n	$\frac{n(n+\beta)(n+\alpha+\beta+N+1)}{(2n+\alpha+\beta)(2n+\alpha+\beta+1)}$
γ_n	$-(b_n+c_n)$
d_n^2	$\frac{(\alpha+\beta+1)(\beta+1)_n(N+\alpha+\beta+2)_n}{\binom{N}{n}(2n+\alpha+\beta+1)(\alpha+1)_n(\alpha+\beta+1)_n}$

Appendix C

Table of Duality Relations Between Classical Orthogonal Polynomials

TABLE C.1. Duality Relations Between Classical Orthogonal Polynomials

Name	$\mathbf{Q_n(x) = \hat{Q}_x(\lambda_n)}$	$\boldsymbol{\lambda_n}$
Hahn–Dual Hahn	$Q_n(x; a, b, N) = R_x(\lambda_n; a, b, N)$	$n(n+a+b+1)$
Meixner–Meixner	$M_n(x; \gamma, \mu) = M_x(\lambda_n; \gamma, \mu)$	n
Krawtchouk–Krawtchouk	$K_n(x; N, p) = K_x(\lambda_n; N, p)$	n
Charlier–Charlier	$C_n(x; \mu) = C_x(\lambda_n; \mu)$	n

Appendix D
Tables of Sheffer Systems

First we list some Sheffer polynomials and their generating functions. Next, in the i.i.d. case we give the characteristic function and distribution of the X_i and for the Lévy case we give the characteristic function and the distribution of X_1. Finally we state some interesting martingale relations. We always set $q = 1 - p$.

D.1 Sheffer Polynomials and Their Generating Functions

TABLE D.1. Sheffer Polynomials and Their Generating Functions

Sheffer Polynomial	**Generating Function**
Bernoulli	$\sum_{m=0}^{\infty} B_m^{(n)}(y)\frac{w^m}{m!} = (w/(e^w-1))^n e^{wy}$
Krawtchouk	$\sum_{m=0}^{\infty} \tilde{K}_m(y;n,p)\frac{w^m}{m!} = (1+(1-p)w)^y(1-pw)^{n-y}$
Euler	$\sum_{m=0}^{\infty} E_m^{(n)}(y)\frac{w^m}{m!} = (2/(1+e^w))^n e^{wy}$
Narumi	$\sum_{m=0}^{\infty} N_m^{(n)}(y)w^m = (\log(1+w)/w)^n(1+w)^y$
Hermite	$\sum_{m=0}^{\infty} \tilde{H}_m(y;t)\frac{w^m}{m!} = \exp(\sqrt{2}yw - w^2t)$
Charlier	$\sum_{m=0}^{\infty} \tilde{C}_m(y;t)\frac{w^m}{m!} = e^{-tw}(1+w)^y$
Laguerre	$\sum_{m=0}^{\infty} \tilde{L}_m^{(t-1)}(y)w^m = (1+w)^{-t}\exp(yw/(w+1))$
Meixner	$\sum_{m=0}^{\infty} \tilde{M}_m(y;t,q)\frac{w^m}{m!} = \left(1-\frac{zq}{q-1}\right)^{-t}\left(\frac{q-1-z}{q-1-cz}\right)^y$
Meix.–Pollac.	$\sum_{m=0}^{\infty} \tilde{P}_m(y;t,\zeta)w^m = (1-\frac{we^{i\zeta}}{2\sin\zeta})^{-t+iy}(1-\frac{we^{-i\zeta}}{2\sin\zeta})^{-t-iy}$
Actuarial	$\sum_{m=0}^{\infty} g_m^{(t)}(y)\frac{w^m}{m!} = \exp(tw + y(1-e^w))$

D.2 Sheffer Polynomials and Their Associated Distributions

TABLE D.2. Sheffer Polynomials and Their Associated Distributions

Sheffer Polyn.	$\phi(\theta)$	**Distribution**
Bernoulli	$(\exp(i\theta) - 1)/(i\theta)$	U(0, 1)
Krawtchouk	$p\exp(i\theta) + q$	B(p)
Euler	$(1 + \exp(i\theta))/2$	B(1/2)
Narumi	$(\exp(i\theta) - 1)/(i\theta)$	U(0, 1)
Hermite	$\exp(-\theta^2/2)$	N(0, 1)
Charlier	$\exp(\exp(-i\theta) - 1)$	P(1)
Laguerre	$(1 - i\theta)^{-1}$	G(1, 1)
Meixner	$p/(1 - q\exp(i\theta))$	Pa(1, p)
Meixner–Pollaczek	$(\cos(a/2)/\cosh((\theta - ia)/2))^2$	
Actuarial	$(1 - i\theta)^{-1}$	G(1, 1)

D.3 Martingale Relations with Sheffer Polynomials

TABLE D.3. Martingale Relations with Sheffer Polynomials

Sheffer Polynomials	**Martingale Equality**
Bernoulli	$E[B_m^{(n)}(S_n) \mid S_k] = B_m^{(k)}(S_k)$
Krawtchouk	$E[\tilde{K}_m(S_n; n, p) \mid S_k] = \tilde{K}_m(S_k; k, p)$
Euler	$E[E_m^{(n)}(S_n) \mid S_k] = E_m^{(k)}(S_k)$
Narumi	$E[N_m^{(n)}(S_n) \mid S_k] = N_m^{(k)}(S_k)$
Hermite	$E[\tilde{H}_m(B_t; t) \mid B_s] = \tilde{H}_m(B_s; s)$
Charlier	$E[\tilde{C}_m(N_t; t) \mid N_s] = \tilde{C}_m(N_s; s)$
Laguerre	$E[\tilde{L}_m^{(t-1)}(G_t) \mid G_s] = \tilde{L}_m^{(s-1)}(G_s)$
Meixner	$E[\tilde{M}_m(P_t; t, q) \mid P_s] = \tilde{M}_m(P_s; s, q)$
Meixner–Pollaczek	$E[\tilde{P}_m(H_t; t, \zeta) \mid H_s] = \tilde{P}_m(H_s; s, \zeta)$
Actuarial	$E[g_m^{(t)}(G_t) \mid G_s] = g_m^{(s)}(G_s)$

Appendix E

Tables of Limit Relations Between Orthogonal Polynomials in the Askey Scheme

TABLE E.1. Limit Relations in the Askey Scheme

Wilson → Continuous Dual Hahn	$\lim_{d\to\infty} \frac{W_n(x^2;a,b,c,d)}{(a+d)_n}$ $= S_n(x^2;a,b,c)$
Wilson → Continuous Hahn	$\lim_{t\to\infty} \frac{W_n((x+t)^2;a-it,b-it,c+it,d+it)}{(-2t)^n n!}$ $= p_n(x;a,b,c,d)$
Wilson → Jacobi	$\lim_{t\to\infty} \frac{W_n(\frac{1-x}{2}t^2;\frac{\alpha+1}{2},\frac{\alpha+1}{2},\frac{\beta+1}{2}+it,\frac{\beta+1}{2}-it)}{t^{2n} n!}$ $= P_n^{(\alpha,\beta)}(x)$
Racah → Hahn	$\lim_{\delta\to\infty} R_n(\lambda(x);\alpha,\beta,-N-1,\delta)$ $= Q_n(x;\alpha,\beta,N)$
Racah → Hahn	$\lim_{\gamma\to\infty} R_n(\lambda(x);\alpha,\beta,\gamma,-\beta-N-1)$ $= Q_n(x;\alpha,\beta,N)$
Racah → Hahn	$\lim_{\delta\to\infty} R_n(\lambda(x);-N-1,\beta+\gamma+N+1,\gamma,\delta)$ $= Q_n(x;\alpha,\beta,N)$
Racah → Dual Hahn	$\lim_{\beta\to\infty} R_n(\lambda(x);-N-1,\beta,\gamma,\delta)$ $= R_n(\lambda(x);\gamma,\delta,N)$

TABLE E.1. (continued) Limit Relations in the Askey Scheme

Racah → Dual Hahn	$\lim_{\alpha\to\infty} R_n(\lambda(x);\alpha,-\delta-N-1,\gamma,\delta)$ $= R_n(\lambda(x);\gamma,\delta,N)$
Racah → Dual Hahn	$\lim_{\beta\to\infty} R_n(\lambda(x);\alpha,\beta,-N-1,$ $\alpha+\delta+N+1) = R_n(\lambda(x);\alpha,\delta,N)$
Cont. Dual Hahn → Meixner–Pollaczek	$\lim_{t\to\infty} \frac{S_n((x-t)^2;\lambda+it,\lambda-it,t\cot(\phi))}{(t/\sin\phi)_n n!}$ $= P_n^{(\lambda)}(x;\phi)$
Continuous Hahn → Meixner–Pollaczek	$\lim_{t\to\infty} \frac{p_n(x-t;\lambda+it,-t\tan(\phi),\lambda-it,-t\tan(\phi))}{(it/\cos\phi)_n i^n}$ $= P_n^{(\lambda)}(x;\phi)$
Continuous Hahn → Jacobi	$\lim_{t\to\infty} \frac{p_n(\frac{-xt}{2};\frac{\alpha+1+it}{2},\frac{\beta+1-it}{2},\frac{\alpha+1-it}{2},\frac{\beta+1+it}{2},)}{(-t)^n}$ $= P_n^{(\alpha,\beta)}(x)$
Continuous Hahn → Jacobi	$\lim_{t\to\infty} \frac{p_n(\frac{-xt}{2};\frac{\alpha+1+it}{2},\frac{\beta+1-it}{2},\frac{\alpha+1-it}{2},\frac{\beta+1+it}{2},)}{(-t)^n}$ $= P_n^{(\alpha,\beta)}(x)$
Hahn → Jacobi	$\lim_{N\to\infty} Q_n(Nx;\alpha,\beta,N)$ $= P_n^{(\alpha,\beta)}(1-x)/P_n^{(\alpha,\beta)}(1)$
Hahn → Meixner	$\lim_{N\to\infty} Q_n(x;b-1,\frac{N(1-c)}{c},N)$ $= M_n(x;b,c)$
Hahn → Krawtchouk	$\lim_{t\to\infty} Q_n(x;pt,(1-p)t,N)$ $= K_n(x;N,p)$
Dual Hahn → Meixner	$\lim_{N\to\infty} R_n(\lambda(x);\beta-1,\frac{N(1-c)}{c},N)$ $= M_n(x;\beta,c)$
Dual Hahn → Krawtchouk	$\lim_{t\to\infty} R_n(\lambda(x);pt,(1-p)t,N) =$ $K_n(x;N,p)$
Meixner–Pollac. → Laguerre	$\lim_{\phi\to 0} P_n^{((\alpha+1)/2)}(-x/(2\phi);\phi)$ $= L_n^{(\alpha)}(x)$
Meixner–Pollac. → Hermite	$\lim_{\lambda\to\infty} \lambda^{-n/2} P_n^{(\lambda)}(\frac{x\sqrt{\lambda}-\lambda\cos\phi}{\sin\phi};\phi)$ $= H_n(x)/n!$

TABLE E.1. (continued) Limit Relations in the Askey Scheme

Jacobi → Laguerre	$\lim_{\beta\to\infty} P_n^{(\alpha,\beta)}(1-\frac{2x}{\beta})$ $= L_n^{(\alpha)}(x)$
Jacobi → Hermite	$\lim_{\alpha\to\infty} \alpha^{-n/2} P_n^{(\alpha,\alpha)}(\frac{x}{\sqrt{\beta}})$ $= H_n(x)/(2^n n!)$
Meixner → Laguerre	$\lim_{c\to 1} M_n(\frac{x}{1-c}; \alpha+1, c)$ $= L_n^{(\alpha)}(x)/L_n^{(\alpha)}(0)$
Meixner → Charlier	$\lim_{\beta\to\infty} M_n(x; \beta, \frac{a}{a+\beta})$ $= C_n(x;a)$
Krawtchouk → Hermite	$\lim_{N\to\infty} \sqrt{\binom{N}{n}} K_n(pN + x\sqrt{2p(1-p)N}; N, p)$ $= (-1)^n H_n(x)/\sqrt{2^n n! \left(\frac{p}{1-p}\right)^n}$
Laguerre → Hermite	$\lim_{\alpha\to\infty} (2/\alpha)^{n/2} L_n^{(\alpha)}(\sqrt{2\alpha}x + \alpha)$ $= (-1)^n H_n(x)/(n!)$
Charlier → Hermite	$\lim_{a\to\infty} (2a)^{n/2} C_n(\sqrt{2a}x + a; a)$ $= (-1)^n H_n(x)$

References

[1] Al-Salam, W.A. and Chihara, T.S. (1976), Convolutions of orthonormal polynomials. *SIAM J. on Math. Anal.* **7** (1), 16–28.

[2] Anderson, W.J. (1991), *Continuous-Time Markov Chains — An Applications-Oriented Approach.* Springer-Verlag, New York.

[3] Andrews, G.E. and Askey, R. (1985), Classical orthogonal polynomials. In *Polynomes Orthogonaux et Applications* (C. Brezinski et al., eds.), Lecture Notes in Mathematics 1171, Springer, Berlin.

[4] Arratia, R., Goldstein, L., and Gordon, L. (1990), Poisson–approximation and the Chen–Stein method. *Statistical Science* **5** (4), 403–434.

[5] Askey, R. and Wilson, J. (1985), *Some basic hypergeometric polynomials that generalize Jacobi polynomials.* Memoirs Amer. Math. Soc. **319**, AMS, Providence RI.

[6] Van Assche, W., Parthasarathy, P.R., and Lenin, R.B. (1999), Spectral representation of certain finite birth and death processes. *The Mathematical Scientist* **24** (2), *to appear.*

[7] Bachelier, L. (1900), Théorie de la spéculation. *Ann. Sci. Ecole Norm. Sup.* **17**, 21–86.

[8] Barbour, A.D. (1980), Equilibrium distributions for Markov population processes. *Adv. Appl. Prob.* **12**, 591–614.

[9] Barbour, A.D. (1988), Stein's method and Poisson process convergence. In *A Celebration of Applied Probability* (J. Gani, ed.), *J. Appl. Prob.* **25A**, 175–184.

[10] Barbour, A.D. (1990), Stein's method for diffusion approximations. *Probab. Theory Related Fields* **84**, 297–322.

[11] Barbour, A.D. and Eagleson, G.K. (1983), Poisson approximation for some statistics based on exchangeable trials. *Adv. Appl. Prob.* **15**, 585–600.

[12] Barbour, A.D., Holst, L., and Janson, S. (1992), *Poisson Approximation.* Clarendon, Oxford.

[13] Barbour, A.D. and Utev, S. (1998), Solving the Stein equation in compound Poisson approximation. *Adv. Appl. Prob.* **30**, 449–475.

[14] Bavinck, H. (1998), Differential operators having Laguerre type and Sobolev type Laguerre orthogonal polynomials as eigenfunctions: A survey. In *Special Functions and Differential Equations, Proceedings Workshop* (Madras), (K. Srinivasa Rao et al., eds.), Allied Publishers, New Delhi.

[15] Bavinck, H. and van Haeringen, H. (1994), Difference equations for generalized Meixner polynomials. *J. Math. Anal Appl.* **184**, 453–463.

[16] Bertoin, J. (1996), *Lévy Processes.* Cambridge University Press, Cambridge.

[17] Bhattacharya, R.N. and Waymire, E.C. (1990), *Stochastic Processes with Applications.* Wiley, New York.

[18] Biane, P. (1990), Chaotic representation for finite Markov chains. *Stochastics and Stochastics Reports* **30**, 61–88.

[19] Billingsley, P. (1995), *Probability and Measure*, 3rd edition. Wiley, New York.

[20] Bingham, N.H. and Kiesel, R. (1998), *Risk-Neutral Valuation. Pricing and Hedging of Financial Derivatives.* Springer-Verlag, Berlin Heidelberg New York.

[21] Black, F. and Scholes, M. (1973), The pricing of options and corporate liabilities. *J. Political Economy* **81**, 635–654.

[22] Boas, R.P.,Jr. and Buck, R.C. (1964), *Polynomial Expansions of Analytic Functions.* Springer-Verlag, Berlin-Göttingen-Heidelberg.

[23] Chen, L.H.Y. (1975), Poisson approximation for dependent trials. *Ann. Probab.* **22**, 1607–1618.

[24] Chen, L.H.Y. (1998), Stein's method: Some perspectives with applications. In *Probability Towards 2000* (L. Accardi and C. C. Heyde, eds.), Lecture Notes in Statistics 128, Springer-Verlag, 97–122.

[25] Chihara, T.S. (1978), *An Introduction to Orthogonal Polynomials.* Gordon and Breach, New York.

[26] Dellacherie, Cl., Maisonneuve, B., and Meyer, P.-A. (1992), *Probabilités et Potentiel*, Herman, Paris, Chapters XVII–XXIV. .

[27] Dermoune, A. (1990), Distribution sur l'espace de P.Lévy et calcul stochastique. *Ann. Inst. Henri. Poincaré* **26** (1), 101–119.

[28] Dette, H. (1994), On a generalization of the Ehrenfest urn model. *J. Appl. Prob.* **31**, 930–939.

[29] Diaconis, P. and Zabell, S. (1991), Closed form summation for classical distributions: Variations on a theme of de Moivre. *Statistical Science* **6**, 284–302.

[30] Di Bucchianico, A. (1997), *Probabilistic and Analytical Aspects of the Umbral Calculus.* CWI Tract **119**. CWI, Amsterdam.

[31] Dickson, D.C.M. and Waters, H.R. (1993), Gamma processes and finite time survival probabilities. *Astin Bulletin* **23** (2), 259–272.

[32] Dickson, D.C.M. and Waters, H.R. (1996), Reinsurance and ruin. *Insurance and Economics* **19**, 61–80.

[33] van Doorn, E.A. (1979), *Stochastic Monotonicity of Birth–Death Processes.* Ph.D. thesis, Technische Hogeschool Twente, Enschede.

[34] van Doorn, E.A. (1980), Stochastic monotonicity of birth–death processes. *Adv. Appl. Prob.* **12**, 59–80.

[35] van Doorn, E.A. (1980), *Stochastic Monotonicity and Queueing Applications of Birth–Death Processes.* Lecture Notes in Statistics 4. Springer-Verlag, New York.

[36] van Doorn, E.A. (1985), Conditions for exponential ergodicity and bounds for the decay parameter of a birth–death process. *Adv. Appl. Prob.* **17**, 514–530.

[37] van Doorn, E.A. (1989), Orthogonal polynomials and birth–death processes. In *Lecture Notes in Pure and Applied Mathematics* Vol. 117, (J. Vinuesa, ed.), Marcel Dekker, New York, 23–34.

[38] van Doorn, E.A. and Schrijner, P. (1993), Random walk polynomials and random walk measures. *J. Comp. Appl. Math.* **49**, 289–296.

[39] Dufresne, F. and Gerber, H.U. (1993), The probability of ruin for the inverse Gaussian and related processes. *Insurance: Mathematics and Economics* **12**, 9–22.

[40] Dufresne, F., Gerber, H.U., and Shiu, E.S.W. (1991), Risk theory and the gamma process. *Astin Bulletin* **22**, 177–192.

[41] Dzhaparidze, K. and van Zuijlen, M. (1996), Introduction to option pricing in a securities market I: Binary models. *CWI Quarterly* **9** (1), 319–356.

[42] Eagleson, G.K. (1968), A duality relation for discrete orthogonal systems. *Studia Scientiarum Mathematicarum Hungarica* **3**, 127–136.

[43] Eberlein, E. and Keller, U. (1995), Hyberbolic distributions in finance. *Bernoulli* **1**, 281–299.

[44] Emery, M. (1989), On the Azéma martingales. In *Lecture Notes in Mathematics* 1372, Springer-Verlag, New York, 66–87.

[45] Engel, D.D. (1982), *The Multiple Stochastic Integral.* Memoires of the American Mathematical Society, 265. AMS, Providence.

[46] Erdélyi, A. (editor) (1955), *Higher Transcendental Functions.* Bateman Manuscript Project, Vol. 3. McGraw-Hill, New York.

[47] Feller, W. (1966), *An Introduction to Probability Theory and Its Applications*, Volume II. Wiley, New York.

[48] Flaspohler, D.C. (1974), Quasi-stationary distributions for absorbing continuous-time denumerable Markov chains. *Ann. Inst. Statist. Math.* **26**, 351–356.

[49] Gerber, H.U. (1992), On the probability of ruin for infinitely divisible claim amount distribution. *Insurance: Mathematics and Economics* **11**, 163–166.

[50] Good, P. (1968), The limiting behavior of transient birth and death processes conditioned on survival. *J. Australian Mathematical Society* **8**, 716–722.

[51] Gradshteyn, I.S. and Ryzhik, I.M. (1980), *Table of Integrals, Series and Products.* Academic Press, London.

[52] Grigelionis, B. (1998), *Processes of Meixner Type.* Matematikos ir Informatikos Institutas Preprintas Nr. 98-18, Vilnius.

[53] Hida, T. (1980), *Brownian Motion.* Springer-Verlag, New York.

[54] Ismail, M.E.H. (1974), Orthogonal polynomials in a certain class of polynomials. *Buletinul Institutului Politehnic Din Iasi* **20** (24), 45–50.

[55] Ismail, M.E.H., Letessier, J., and Valent, G. (1988), Linear birth and death models and associated Laguerre and Meixner polynomials. *J. Approximation Theory* **55**, 337–348.

[56] Ito, K. (1951), Multiple Wiener integral. *J. Mathematical Society of Japan* **3** (1), 157–169.

[57] Ito, Y. (1988), Generalized Poisson functionals. *Probab. Th. Rel. Fields* **77**,1–28.

[58] Johnson, N.L., Kotz, S., and Kemp, A.W. (1992), *Univariate Discrete Distributions*, 2nd edition. Wiley, New York.

[59] Karlin, S. and McGregor, J.L. (1957), The differential equations of birth-and-death processes, and the Stieltjes moment problem. *Trans. Amer. Math. Soc.* **85**, 489–546.

[60] Karlin, S. and McGregor, J.L. (1957), The classification of birth and death processes. *Trans. Amer. Math. Soc.* **86**, 366–400.

[61] Karlin, S. and McGregor, J.L. (1958), Linear growth, birth and death processes. *J. Math. Mech.* **7**, 643–662.

[62] Karlin, S. and McGregor, J.L. (1959), Random walks. *Illinois J. Math.* **8**, 87–118.

[63] Karlin, S. and McGregor, J.L. (1965), Ehrenfest urn models. *J. Appl. Prob.* **19**, 477–487.

[64] Karlin, S. and Taylor, H.M. (1975), *A First Course in Stochastic Processes*, 2nd edition. Academic Press, New York.

[65] Karlin, S. and Taylor, H.M. (1975), *A Second Course in Stochastic Processes.* Academic Press, New York.

[66] Kendall, D.G. (1959), Unitary dilations of one-parameter semigroups of Markov transition operators, and the corresponding integral representation for Markov processes with a countable infinity of states. *Proc. London Math. Soc.* **3** (9), 417–431.

[67] Kendall, D.G. and Reuter, G.E.H. (1959), The calculation of the ergodic projection for Markov chains and processes with a countable infinity of states. *Acta Math.* **97**, 103–144.

[68] Kijima, M., Nair, M.G, Pollett, P.K., and van Doorn, E.A. (1997), Limiting conditional distributions for birth–death processes. *Adv. Appl. Prob.* **29**, 185–204.

[69] Knuth, D.E. (1981), *The Art of Computer Programming, Vol. I: Fundamental Algorithms.* Addison-Wesley, Readings, Mass.

[70] Koekoek, R. and Swarttouw, R.F. (1994), *The Askey-scheme of hypergeometric orthogonal polynomials and its q-analogue.* Report 94-05, Delft University of Technology.

[71] Koekoek, R. and Swarttouw, R.F. (1998), *The Askey-scheme of hypergeometric orthogonal polynomials and its q-analogue.* Report 98-17, Delft University of Technology.

[72] Koornwinder, T.H. (1994), *q-Special Functions, a Tutorial.* Report 94-08, Department of Mathematics and Computer Science, University of Amsterdam.

[73] Labelle, J. (1985), *Polynomes Orthogonaux et Applications,* (C. Brezinski et al., eds.), Lecture Notes in Mathematics 1171, Springer, Berlin, xxxvii.

[74] Ledermann, W. and Reuter, G.E.H. (1954), Spectral theory for the differential equations of simple birth and death processes. *Philos. Trans. Roy. Soc. London, Ser. A* **246**, 321–369.

[75] Lesky, P.A. (1995), Vervollständigung der klassischen Orthogonalpolynome durch Ergänzungen zum Askey-Schema der hypergeometrische orthogonalen Polynome. *Sitzungsber. Abt. II* **204**, 151–166.

[76] L. Littlejohn (1986), An application of a new theorem on orthogonal polynomials and differential equations. *Quaestiones Math.* **10**, 49–61.

[77] Luk, H.M. (1994), *Stein's Method for the Gamma Distribution and Related Statistical Applications.* Ph.D.Thesis. University of Southern California, Los Angeles.

[78] Lukacs, E. (1970), *Characteristic Functions.* Griffin, London.

[79] Madan, D.B. and Seneta, E. (1990), The variance gamma (V.G.) model for share market returns. *J. of Business* **63** (4), 511–524.

[80] Mazet, O. (1997), Classification des semi-groupes de diffusion sur **R** associés à une famille de polynômes orthogonaux. In *Séminaire de Probabilités XXXI,* Lectures Notes in Mathematics 1655, (J. Azéma, M. Emery, and M. Yor, eds.), Springer-Verlag, Berlin, 40–53.

[81] Meixner, J. (1934), Orthogonale Polynomsysteme mit einer besonderen Gestalt der erzeugende Funktion. *J. London Math. Soc.* **9**, 6–13.

[82] Meyer, P.A. (1986), *Eléments de Probabilités Quantiques, Séminaire de Probabilités no. XX,* Lecture Notes in Mathematics 1024, Springer-Verlag, Heidelberg.

[83] Neveu, J. (1975), *Discrete-Parameter Martingales.* North-Holland.

[84] Nikiforov, A.F., Suslov, S.K., and Uvarov, V.B. (1991), *Classical Orthogonal Polynomials of a Discrete Variable.* Springer-Verlag, Berlin.

[85] van Noortwijk, J. (1996), *Optimal Maintenance Decisions for Hydraulic Structures Under Isotropic Deterioration.* Ph.D. thesis, Delft University of Technology.

[86] Nualart, D. and Schoutens, W. (1999), *Chaotic and Predictable Representations for Lévy Processes*, Mathematics Preprint Series No. 269. Universitat de Barcelona.

[87] Ogora, H. (1972), Orthogonal functionals of the Poisson process. *IEEE Transactions on Information Theory* **IT-18** (4), 474–481.

[88] Parthasarathy, P.R., Lenin, R.B., Schoutens, W., and Van Assche, W. (1998), A birth and death process related to the Rogers–Ramanujan continued fraction. *J. Math. Anal. Appl.* **224**, 297–315.

[89] Plucińska, A. (1998), A stochastic characterization of Hermite polynomials. *J. of Mathematical Sciences* **89** (5), 1541–1544.

[90] Plucińska, A. (1998), Polynomial normal densities generated by Hermite polynomials. *J. of Mathematical Sciences* **92** (3), 3921–3925.

[91] Pommeret, D. (2000), Orthogonality of the Sheffer system associated to a Lévy process. *J. of Statistical Planning and Interference, to appear.*

[92] Privault, N. (1994), Chaotic and variational calculus in discrete and continuous time for the Poisson process. *Stochastics and Stochastics Reports* **51**, 83–109.

[93] Privault, N., Solé, J.L., and Vives, J. (2000), Chaotic Kabanov formula for the Azéma martingales. *Bernoulli, to appear.*

[94] Protter, P. (1990), *Stochastic Integration and Differential Equations.* Springer-Verlag, Berlin.

[95] Reinert, G. (1998), Stein's method and applications to empirical measures. *Aportaciones Matemáticas, Modelos Estocásticos* **14**, 65–120.

[96] Roberts, G.O. and Jacka, S.D. (1994), Weak convergence of conditioned birth and death processes. *J. Appl. Prob.* **31**, 90–100.

[97] Rolski, T., Schmidli, H., Schmidt, V., and Teugels, J. (1999) *Stochastic Processes for Insurance and Finance.* Wiley, Chichester.

[98] Sato, K. (1999), *Lévy Processes and Infinitely Divisible Distributions.* Cambridge University Press, Cambridge.

[99] Schoutens, W. (1998), Lévy–Sheffer and IID-Sheffer polynomials with applications to stochastic integrals. *J. Comp. Appl. Math.* **99** (1 and 2), 365–372.

[100] Schoutens, W. (1998), Integration and summation formulas for Sheffer polynomials based on martingale relations. *Aportaciones Matemáticas, Modelos Estocásticos* **14**, 317–324.

[101] Schoutens, W. (1999), *Stochastic Processes in the Askey Scheme.* Doctoral dissertation, K. U. Leuven.

[102] Schoutens, W. (1999), *Orthogonal Polynomials in Stein's Method,* EURANDOM Report No. 99-041, EURANDOM, Eindhoven.

[103] Schoutens, W. and Teugels, J.L. (1998), Lévy processes, polynomials and martingales. *Commun. Statist. — Stochastic Models* **14** (1 and 2), 335–349.

[104] Schrijner, P. (1995), *Quasi-Stationarity of Discrete-Time Markov Chains.* Ph.D. thesis, Universiteit Twente, Enschede.

[105] Schrijner, P. and van Doorn, E.A. (1997), Weak convergence of conditioned birth-death processes in discrete time. *J. Appl. Prob.* **34**, 46–53.

[106] Sheffer, I.M. (1937), Concerning Appell sets and associated linear functional equations. *Duke Math. J.* **3**, 593–609.

[107] Sheffer, I.M. (1939), Some properties of polynomial sets of type zero. *Duke Math. J.* **5**, 590–622.

[108] Stein, C. (1972), A bound for the error in the normal approximation to the distribution of a sum of dependent random variables. *Proc. Sixth Berkeley Symp. Math. Statist. Probab.* **2**, 583–602. University of California Press, Berkeley.

[109] Stein, C. (1986), *Approximate Computation of Expectations.* IMS Lecture Notes — Monograph Series, Vol. 7. IMS, Hayward, Calif.

[110] Szegö, G. (1939), *Orthogonal Polynomials.* AMS Colloquium Publications, Vol. 23. AMS, Providence, R.I.

[111] Wesołowski, J. (1990), A martingale characterization of the Poisson process. *Bull. Polish Acad. Sci. Math.* **38**, 49–53.

[112] Wiener, N. (1930), The homogeneous chaos. *Amer. J. Math.* **60**, 897–936.

[113] Williams, D. (1991), *Probability with Martingales.* Cambridge University Press, London.

[114] Wilson, J.A. (1980), Some hypergeometric orthogonal polynomials. *SIAM J. Math. Anal.* 11, 690–701.

Index

Lecture Notes in Statistics

For information about Volumes 1 to 72, please contact Springer-Verlag

Vol. 73: B.C. Arnold, E. Castillo, J.-M. Sarabia, Conditionally Specified Distributions. xiii, 151 pages, 1992.

Vol. 74: P. Barone, A. Frigessi, M. Piccioni, Stochastic Models, Statistical Methods, and Algorithms in Image Analysis. vi, 258 pages, 1992.

Vol. 75: P.K. Goel, N.S. Iyengar (Eds.), Bayesian Analysis in Statistics and Econometrics. xi, 410 pages, 1992.

Vol. 76: L. Bondesson, Generalized Gamma Convolutions and Related Classes of Distributions and Densities. viii, 173 pages, 1992.

Vol. 77: E. Mammen, When Does Bootstrap Work? Asymptotic Results and Simulations. vi, 196 pages, 1992.

Vol. 78: L. Fahrmeir, B. Francis, R. Gilchrist, G. Tutz (Eds.), Advances in GLIM and Statistical Modelling: Proceedings of the GLIM92 Conference and the 7th International Workshop on Statistical Modelling, Munich, 13-17 July 1992. ix, 225 pages, 1992.

Vol. 79: N. Schmitz, Optimal Sequentially Planned Decision Procedures. xii, 209 pages, 1992.

Vol. 80: M. Fligner, J. Verducci (Eds.), Probability Models and Statistical Analyses for Ranking Data. xxii, 306 pages, 1992.

Vol. 81: P. Spirtes, C. Glymour, R. Scheines, Causation, Prediction, and Search. xxiii, 526 pages, 1993.

Vol. 82: A. Korostelev and A. Tsybakov, Minimax Theory of Image Reconstruction. xii, 268 pages, 1993.

Vol. 83: C. Gatsonis, J. Hodges, R. Kass, N. Singpurwalla (Editors), Case Studies in Bayesian Statistics. xii, 437 pages, 1993.

Vol. 84: S. Yamada, Pivotal Measures in Statistical Experiments and Sufficiency. vii, 129 pages, 1994.

Vol. 85: P. Doukhan, Mixing: Properties and Examples. xi, 142 pages, 1994.

Vol. 86: W. Vach, Logistic Regression with Missing Values in the Covariates. xi, 139 pages, 1994.

Vol. 87: J. Müller, Lectures on Random Voronoi Tessellations.vii, 134 pages, 1994.

Vol. 88: J. E. Kolassa, Series Approximation Methods in Statistics. Second Edition, ix, 183 pages, 1997.

Vol. 89: P. Cheeseman, R.W. Oldford (Editors), Selecting Models From Data: AI and Statistics IV. xii, 487 pages, 1994.

Vol. 90: A. Csenki, Dependability for Systems with a Partitioned State Space: Markov and Semi-Markov Theory and Computational Implementation. x, 241 pages, 1994.

Vol. 91: J.D. Malley, Statistical Applications of Jordan Algebras. viii, 101 pages, 1994.

Vol. 92: M. Eerola, Probabilistic Causality in Longitudinal Studies. vii, 133 pages, 1994.

Vol. 93: Bernard Van Cutsem (Editor), Classification and Dissimilarity Analysis. xiv, 238 pages, 1994.

Vol. 94: Jane F. Gentleman and G.A. Whitmore (Editors), Case Studies in Data Analysis. viii, 262 pages, 1994.

Vol. 95: Shelemyahu Zacks, Stochastic Visibility in Random Fields. x, 175 pages, 1994.

Vol. 96: Ibrahim Rahimov, Random Sums and Branching Stochastic Processes. viii, 195 pages, 1995.

Vol. 97: R. Szekli, Stochastic Ordering and Dependence in Applied Probability. viii, 194 pages, 1995.

Vol. 98: Philippe Barbe and Patrice Bertail, The Weighted Bootstrap. viii, 230 pages, 1995.

Vol. 99: C.C. Heyde (Editor), Branching Processes: Proceedings of the First World Congress. viii, 185 pages, 1995.

Vol. 100: Wlodzimierz Bryc, The Normal Distribution: Characterizations with Applications. viii, 139 pages, 1995.

Vol. 101: H.H. Andersen, M.Højbjerre, D. Sørensen, P.S.Eriksen, Linear and Graphical Models: for the Multivariate Complex Normal Distribution. x, 184 pages, 1995.

Vol. 102: A.M. Mathai, Serge B. Provost, Takesi Hayakawa, Bilinear Forms and Zonal Polynomials. x, 378 pages, 1995.

Vol. 103: Anestis Antoniadis and Georges Oppenheim (Editors), Wavelets and Statistics. vi, 411 pages, 1995.

Vol. 104: Gilg U.H. Seeber, Brian J. Francis, Reinhold Hatzinger, Gabriele Steckel-Berger (Editors), Statistical Modelling: 10th International Workshop, Innsbruck, July 10-14th 1995. x, 327 pages, 1995.

Vol. 105: Constantine Gatsonis, James S. Hodges, Robert E. Kass, Nozer D. Singpurwalla(Editors), Case Studies in Bayesian Statistics, Volume II. x, 354 pages, 1995.

Vol. 106: Harald Niederreiter, Peter Jau-Shyong Shiue (Editors), Monte Carlo and Quasi-Monte Carlo Methods in Scientific Computing. xiv, 372 pages, 1995.

Vol. 107: Masafumi Akahira, Kei Takeuchi, Non-Regular Statistical Estimation. vii, 183 pages, 1995.

Vol. 108: Wesley L. Schaible (Editor), Indirect Estimators in U.S. Federal Programs. viii, 195 pages, 1995.

Vol. 109: Helmut Rieder (Editor), Robust Statistics, Data Analysis, and Computer Intensive Methods. xiv, 427 pages, 1996.

Vol. 110: D. Bosq, Nonparametric Statistics for Stochastic Processes. xii, 169 pages, 1996.

Vol. 111: Leon Willenborg, Ton de Waal, Statistical Disclosure Control in Practice. xiv, 152 pages, 1996.

Vol. 112: Doug Fischer, Hans-J. Lenz (Editors), Learning from Data. xii, 450 pages, 1996.

Vol. 113: Rainer Schwabe, Optimum Designs for Multi-Factor Models. viii, 124 pages, 1996.

Vol. 114: C.C. Heyde, Yu. V. Prohorov, R. Pyke, and S. T. Rachev (Editors), Athens Conference on Applied Probability and Time Series Analysis Volume I: Applied Probability In Honor of J.M. Gani. viii, 424 pages, 1996.

Vol. 115: P.M. Robinson, M. Rosenblatt (Editors), Athens Conference on Applied Probability and Time Series Analysis Volume II: Time Series Analysis In Memory of E.J. Hannan. viii, 448 pages, 1996.

Vol. 116: Genshiro Kitagawa and Will Gersch, Smoothness Priors Analysis of Time Series. x, 261 pages, 1996.

Vol. 117: Paul Glasserman, Karl Sigman, David D. Yao (Editors), Stochastic Networks. xii, 298, 1996.

Vol. 118: Radford M. Neal, Bayesian Learning for Neural Networks. xv, 183, 1996.

Vol. 119: Masanao Aoki, Arthur M. Havenner, Applications of Computer Aided Time Series Modeling. ix, 329 pages, 1997.

Vol. 120: Maia Berkane, Latent Variable Modeling and Applications to Causality. vi, 288 pages, 1997.

Vol. 121: Constantine Gatsonis, James S. Hodges, Robert E. Kass, Robert McCulloch, Peter Rossi, Nozer D. Singpurwalla (Editors), Case Studies in Bayesian Statistics, Volume III. xvi, 487 pages, 1997.

Vol. 122: Timothy G. Gregoire, David R. Brillinger, Peter J. Diggle, Estelle Russek-Cohen, William G. Warren, Russell D. Wolfinger (Editors), Modeling Longitudinal and Spatially Correlated Data. x, 402 pages, 1997.

Vol. 123: D. Y. Lin and T. R. Fleming (Editors), Proceedings of the First Seattle Symposium in Biostatistics: Survival Analysis. xiii, 308 pages, 1997.

Vol. 124: Christine H. Müller, Robust Planning and Analysis of Experiments. x, 234 pages, 1997.

Vol. 125: Valerii V. Fedorov and Peter Hackl, Model-oriented Design of Experiments. viii, 117 pages, 1997.

Vol. 126: Geert Verbeke and Geert Molenberghs, Linear Mixed Models in Practice: A SAS-Oriented Approach. xiii, 306 pages, 1997.

Vol. 127: Harald Niederreiter, Peter Hellekalek, Gerhard Larcher, and Peter Zinterhof (Editors), Monte Carlo and Quasi-Monte Carlo Methods 1996, xii, 448 pages, 1997.

Vol. 128: L. Accardi and C.C. Heyde (Editors), Probability Towards 2000, x, 356 pages, 1998.

Vol. 129: Wolfgang Härdle, Gerard Kerkyacharian, Dominique Picard, and Alexander Tsybakov, Wavelets, Approximation, and Statistical Applications, xvi, 265 pages, 1998.

Vol. 130: Bo-Cheng Wei, Exponential Family Nonlinear Models, ix, 240 pages, 1998.

Vol. 131: Joel L. Horowitz, Semiparametric Methods in Econometrics, ix, 204 pages, 1998.

Vol. 132: Douglas Nychka, Walter W. Piegorsch, and Lawrence H. Cox (Editors), Case Studies in Environmental Statistics, viii, 200 pages, 1998.

Vol. 133: Dipak Dey, Peter Müller, and Debajyoti Sinha (Editors), Practical Nonparametric and Semiparametric Bayesian Statistics, xv, 408 pages, 1998.

Vol. 134: Yu. A. Kutoyants, Statistical Inference For Spatial Poisson Processes, vii, 284 pages, 1998.

Vol. 135: Christian P. Robert, Discretization and MCMC Convergence Assessment, x, 192 pages, 1998.

Vol. 136: Gregory C. Reinsel, Raja P. Velu, Multivariate Reduced-Rank Regression, xiii, 272 pages, 1998.

Vol. 137: V. Seshadri, The Inverse Gaussian Distribution: Statistical Theory and Applications, xi, 360 pages, 1998.

Vol. 138: Peter Hellekalek, Gerhard Larcher (Editors), Random and Quasi-Random Point Sets, xi, 352 pages, 1998.

Vol. 139: Roger B. Nelsen, An Introduction to Copulas, xi, 232 pages, 1999.

Vol. 140: Constantine Gatsonis, Robert E. Kass, Bradley Carlin, Alicia Carriquiry, Andrew Gelman, Isabella Verdinelli, Mike West (Editors), Case Studies in Bayesian Statistics, Volume IV, xvi, 456 pages, 1999.

Vol. 141: Peter Müller, Brani Vidakovic (Editors), Bayesian Inference in Wavelet Based Models, xi, 394 pages, 1999.

Vol. 142: György Terdik, Bilinear Stochastic Models and Related Problems of Nonlinear Time Series Analysis: A Frequency Domain Approach, xi, 258 pages, 1999.

Vol. 143: Russell Barton, Graphical Methods for the Design of Experiments, x, 208 pages, 1999.

Vol. 144: L. Mark Berliner, Douglas Nychka, and Timothy Hoar (Editors), Case Studies in Statistics and the Atmospheric Sciences, x, 208 pages, 1999.

Vol. 145: James H. Matis and Thomas R. Kiffe, Stochastic Population Models, viii, 220 pages, 2000.

Vol. 146: Wim Schoutens, Stochastic Processes and Orthogonal Polynomials, xiv, 163 pages, 2000.

9 780387 950150